Collins

The Shang Maths Project

For the English National Curriculum

Homework Guide

Author: Peter Lewis-Cole

Homework Guide Series Editor: Amanda Simpson

Practice Books Series Editor: Professor Lianghuo Fan

William Collins' dream of knowledge for all began with the publication of his first book in 1819.

A self-educated mill worker, he not only enriched millions of lives, but also founded a flourishing publishing house. Today, staying true to this spirit, Collins books are packed with inspiration, innovation and practical expertise. They place you at the centre of a world of possibility and give you exactly what you need to explore it.

Collins. Freedom to teach.

Published by Collins
An imprint of HarperCollins*Publishers*
The News Building
1 London Bridge Street
London
SE1 9GF

Browse the complete Collins catalogue at
www.collins.co.uk

© HarperCollins*Publishers* Limited 2019

10 9 8 7 6 5 4 3 2 1

978-0-00-824142-1

Author: Peter Lewis-Cole

Homework Guide Series Editor: Amanda Simpson

Practice Books Series Editor: Professor Lianghuo Fan

British Library Cataloguing in Publication Data

A catalogue record for this publication is available from the British Library.

Publisher: Elizabeth Catford
In-house Senior Editor: Mike Appleton
Project Manager: Emily Hooton
Copy Editor: Karen Williams
Proofreader: Catherine Dakin
Answers: Steven Matchett
Cover design: Kevin Robbins and East China Normal University Press Ltd.
Internal design: 2Hoots Publishing Services Ltd
Typesetting: 2Hoots Publishing Services Ltd
Production: Sarah Burke

Printed and bound by CPI Group (UK) Ltd, Croydon, CR0 4YY

MIX
Paper from
responsible sources

FSC
www.fsc.org **FSC™ C007454**

This book is produced from independently certified FSC paper to ensure responsible forest management.

For more information visit:
www.harpercollins.co.uk/green

Contents

Chapter 9 Geometry and measurement (II)

Chapter 10 Four operations of whole numbers

1.1 Warm up revision

1. Calculate.

(a) $430 + 170 =$ ☐

(b) $427 + 163 =$ ☐

(c) $389 + 112 =$ ☐

(d) $389 + 212 =$ ☐

(e) $579 + 122 - 100 =$ ☐

(f) $579 - 122 + 100 =$ ☐

2. Calculate and find the pattern.

(a) $122 + 888 =$ ☐

$233 + 777 =$ ☐

☐ $+ 444 =$ ☐

$555 + 555 =$ ☐

(b) $374 + 442 =$ ☐

$521 + 394 =$ ☐

$684 + 234 =$ ☐

$755 +$ ☐ $= 919$

(c) $546 - 137 =$ ☐

$728 - 219 =$ ☐

$592 - 343 =$ ☐

$847 -$ ☐ $= 619$

What patterns did you notice?

 Patterns

Together, look at the problems completed above and create two lists of five number sentences that follow similar patterns.

Recall multiplication facts and division facts

1. Complete this table using your multiplication facts knowledge.

×	7	12	8	5	6	9	11	4	10
6				30					60
7				35					70
8				40					80
9				45					90
12				60					120

(a) What do you notice about the 5× and 10× number facts given to you?

(b) Can you find any other sets of multiplication facts where the same thing would happen?

2. Calculate.

   (a) $6 \times 9 = \boxed{}$

   (b) $9 \times 7 = \boxed{}$

   (c) $8 \times \boxed{} = 56$

   (d) $56 = 7 \times \boxed{}$

   (e) $\boxed{} \div 9 = 8$

3. Use your number facts knowledge to solve these.

   (a) $7 \times 15 = 7 \times 10 + 7 \times 5 = \boxed{}$

   (b) $9 \times 13 = 9 \times 10 + 9 \times 3 = \boxed{}$

   (c) $8 \times 14 = \boxed{} + \boxed{} = 112$

👪 Fluency of multiplication facts

Ask your child to write down all of the tables facts up to 12 × 12 that they are currently learning and/or are finding a little tricky to recall automatically, on small pieces of paper. Shuffle the pieces of paper and select from the top each time. Say the number sentence aloud, including the answer. Repeat until your child is confident with all of the facts.

1.3 Multiplication and division (1)

Multiply and divide 3-digit numbers by a single-digit number

1. Calculate the speed of each racer around the track.

Racer	Distance (m)	Time (s)	Speed (m/s)
A	144	9	
B	126	9	
C		8	16
D	126		9

2. Use the column method to calculate these.

(a) 76 × 6 = [ ]

(b) 83 × 7 = [ ]

(c) 67 × 8 = [ ]

(d) 455 ÷ 7 = [ ]

(e) 704 ÷ 8 = [ ]

(f) 657 ÷ [ ] = 73

 **Total cost**

Present this problem to your child: *Sam buys his favourite magazine each month for a whole year. He loves to read all of the 26 pages in the magazine. How many pages will he have read in total in a year?* Ask your child to calculate the answer.

Say: *One day, Sam's mum tells him that he has already read 234 pages. How many magazines has he read?* Ask your child to calculate the answer.

1.4 Multiplication and division (2)

Multiply and divide 3-digit numbers by a single-digit number

1. Use the column method to calculate.

   (a) What is the product if the factors are 7 and 298?

   (b) The product is 1107 and the factor is 9, but what is the other factor?

   (c) The quotient is 406 and the divisor is 6, but what is the dividend?

   (d) The dividend is 4256 and the divisor is 7. What is the quotient?

2. Complete the table.

   In a supermarket, boxes of fruit are displayed. Write how many boxes and the total number of each fruit there are.

Item	Number of fruits in each box	Number of boxes	Total items
Apples	9	110	990
Strawberries	7	280	
Bananas	6		906
Pears	8	399	
Melons	5		1,210

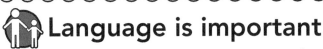 **Language is important**

Ask your child to explain the terms 'product', 'factor', 'quotient', 'dividend' and 'divisor' to you. Can they use diagrams or drawings to help show how they are connected?

1.5 Problem solving (1)

Use strategies to solve multiplication and division problems

1. Use the line model to help you solve this problem.

   Candle B burns 9 times as quickly as candle A.
   Candle A takes 792 minutes to burn.
   How long will candle B burn for?

A   B

2. Show two different methods that would solve these problems.
   (a) Mike and Anna are writing their own novels. Each day, Anna manages to write 4 times as many words as Mike. If Anna writes 1260 words in a day, how many words does Mike write?

Method 1	Method 2

   (b) In a race, Kayo and Finn run 152 metres each minute. Finn completes the race in 7 minutes but Kayo stops after 5 minutes with an injury. How many metres did each person run?

Method 1	Method 2

 Line model

These three numbers have a multiplicative relationship: 15, 195 and 2925.

Create a line model with your child to show the relationship between these numbers. Use the line model to explore the calculations that it shows.

©HarperCollinsPublishers 2019

1.6 Problem solving (2)

Use strategies to solve multiplication and division problems

1. If A = 460, write the number sentences and find the answers.
   (a) A is 4 times as many as B. What is the sum of A and B?

   Number sentence: _____

   Answer: [ ]

   (b) C is 3 times as many as A. What is half the sum of A and C?

   Number sentence: _____

   Answer: [ ]

   (c) Subtract A and B from C. What answer do you get if you divide the answer by 5?

   Number sentence: _____

   Answer: [ ]

2. Use the line models below to solve each problem.
   (a) If Anna has 23 toys, how many toys does Ben have? [ ]

   Anna's toys [▬]
   Ben's toys [▬][▬][▬]

   (b) Fill in the gaps below.
   Farmer A = 32

   Farmer B = [ ]

   Farmer C = [ ]

   Farmer A [▬]
   Farmer B [▬][▬]
   Farmer C [▬][▬][▬]

   Total number of animals = [ ]

👪 Florist mathematics

Work with your child to create a mathematical image that helps to represent this problem.

daisies = 3 times the number of tulips

roses = 4 times the number of sunflowers

tulips = 2 times smaller than the number of lilies

lilies = a third of the number of daisies

sunflowers = double the number of lilies

1.7 Fractions

Recognise, find and calculate fractions of shapes and quantities

1. Write the fraction of the shape that is shaded.

(a)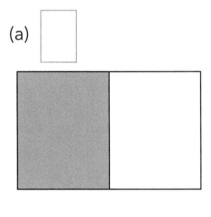

(b)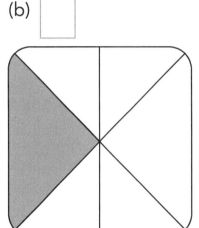

(c)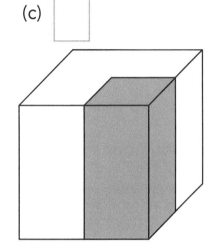

2. Who gets the most cookies?

Emma has $\frac{1}{4}$ of the cookies.

Paul has $\frac{4}{16}$ of the cookies

Mike has $\frac{2}{8}$ of the cookies

George has $\frac{5}{20}$ of the cookies.

_____

👪 Is it possible?

Look at the shapes below. Ask your child how they would represent the following fractions using these shapes. Is it possible? $\frac{1}{2}, \frac{2}{4}, \frac{1}{3}, \frac{4}{5}, \frac{3}{7}$

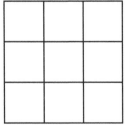 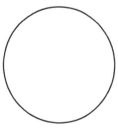  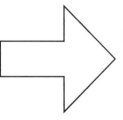 

Place value of 4-digit and 5-digit numbers

1. Complete the table.

Number	Number in words
	Five thousand and five
	Eleven thousand six hundred and forty
28 702	
10 004	
	Twelve thousand and thirty-seven
14 321	

2. Answer the questions using the clues given.

(a) The sum of all the digits in this 4-digit number is 24.

Smallest possible number: ☐

Greatest possible number: ☐

(b) In a 4-digit number, each digit is double or half of another. Can you find four different 4-digit numbers?

☐ ☐ ☐ ☐

(c) Using the digits 5, 7, 0 and 9, find four different numbers that are all multiples of 10.

☐ ☐ ☐ ☐

~~~~~~~~~~~~~~~~~~~~

👪 Numbers are everywhere

Think about numbers that are all around us in our everyday life. Ask your child to make a list. What is the biggest number they can think of? Where can they find it in everyday life? Can they find a 5- or 6-digit number?

Compare and order 4-digit numbers

1. Order these numbers, from largest to smallest.

   6893, 8693, 9836, 6398, 8396

   (a) Are these statements true (✓) or false (✗) about the ordered numbers above?

      (i)   The largest number has 8 thousands. . . . . ☐

      (ii) The smallest number has 39 tens. . . . . ☐

      (iii) A number with 86 hundreds could not be the largest. . . . ☐

      (iv) The smallest number must have 3 ones. . . . . ☐

      (v)   Using the same four digits as these numbers,
            there is a smallest possible number. . . . . ☐

2. Using only the digits given, complete the following problems.

   3, 4, 5, 7, 9, 0

   (a) The smallest possible 4-digit number. ☐

   (b) The largest possible 4-digit number. ☐

   (c) A number that cannot be the smallest or largest. ☐

   (d) A number that has more than 5 thousands. ☐

   (e) A 4-digit number where the sum of the digits is > 20 but < 25. ☐

👪 Number strings

Help your child to create their own number patterns using the > and < symbols to express the relationships between them. Encourage them to create number strings that are 3, 4, 5 and 6 numbers long.

For example:

7362 > 6198 < 6987

Round numbers to the nearest 10, 100 and 1000

1. Write the multiples of a thousand, a hundred and ten nearest the numbers below. Use the number line to help you.

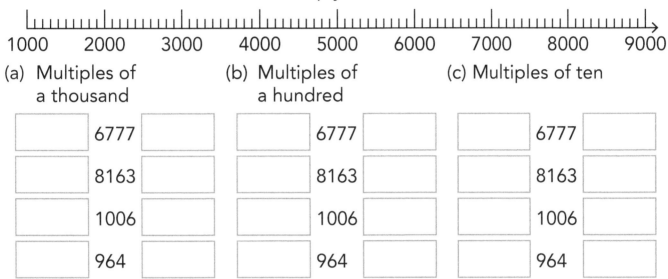

(a) Multiples of a thousand

(b) Multiples of a hundred

(c) Multiples of ten

	6777				6777				6777	
8163				8163				8163	
1006				1006				1006	
964				964				964	

Tick the number that is the closest to the number given.

2. Complete the table. The first one is done for you.

Number	Nearest thousand	Nearest hundred	Nearest ten
7652	8000	7700	7650
5946			
9040			
	4000	4300	4270

3. Write the letter of the town that has the greatest population when rounded in the following ways.

Population of Town A is 4525.   Population of Town B is 4483.

(a) To the nearest thousand = Town ___ has the greatest population.

(b) To the nearest hundred = Town ___ has the greatest population.

Each digit counts

Complete and discuss the following together. Say to your child:

My number is 3412. I think that it doesn't matter to which value I round this number, it will always round down. Do you agree? Can you explain why?

Can you think of a different number where this would also work? Can you think of a number where you would only round up?

©HarperCollinsPublishers 2019

2.4 Addition with 4-digit numbers (1)

Add numbers with up to 4 digits

1. Calculate the following. You can choose which method you use. Use the space given to show your working out.

(a) 3572 + 2318 = [   ]

(b) 5068 + 1055 = [   ]

(c) 6952 + 3048 = [   ]

(d) 5608 + 4356 = [   ]

2. Write the number sentence and calculate.
   (a) What is 1234 more than 5678?

   Answer: _____

   (b) One addend is 5826 and the other addend is 3392. What is the sum?

   Answer: _____

   (c) One addend is 6352 and the sum is 7452, but what is the other addend?

   Answer: _____

👪 Line modelling

Ask your child to draw a line model to represent the relationship between these three numbers: 572, 222, 800. Using the completed line model, they should identify and write the addition and subtraction facts shown, and for each, label the numbers that represent the addend, sum, minuend, subtrahend or difference.

2.5 Addition with 4-digit numbers (2)

Add numbers with up to 4-digits

1. Use the column method to calculate.

    (a) 6107 + 4884 = ◻

    (b) 5852 + 4148 = ◻

    (c) 1804 + 2909 = ◻

    (d) 7568 + 4675 = ◻

2. Fill in the blanks to make these calculations correct. Show any digits that should be under the bottom line.

    (a)
    ```
        8  ◻  ◻  1
      + ◻  3  9  ◻
      ─────────────
        9  8  2  0
    ```

    (b)
    ```
        1  ◻  3  ◻
      + ◻  6  ◻  8
      ─────────────
        6  9  1  2
    ```

    (c)
    ```
        ◻  6  ◻  8
      + 5  ◻  7  ◻
      ─────────────
      1 1  3  5  6
    ```

    (d)
    ```
        9  2  1  4
      +    8  ◻  5
      ─────────────
      1 0  ◻  9  ◻
    ```

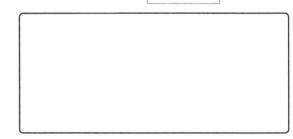 Challenge time

Using the digits 1, 2, 3, 4, 5, 6, 7, 8 and 9 only once, challenge your child to create an addition of two 3-digit numbers. It should look something like this:

```
      ◻  ◻  ◻
  +   ◻  ◻  ◻
  ───────────
   ◻  ◻  ◻
```

2.6 Subtraction with 4-digit numbers (1)

Subtract numbers with up to 4 digits

1. Calculate the following. You can choose which method you use.
   Use the space given to show your working out.

   (a) 6583 – 3472 = ⬚

   (b) 9304 – 4021 = ⬚

   (c) 7005 – 2350 = ⬚

   (d) 8351 – 4257 = ⬚

2. Write the number sentence and calculate.
   (a) What is 4321 less than 8909?

   Answer: _____

   (b) The minuend is 7052 and the subtrahend is 3507. What is the difference?

   Answer: _____

   (c) The subtrahend is 2604 and the difference is 6902, but what is the minuend?

   Answer: _____

 ## Shopping money

On your next visit to the supermarket, let your child find an item for sale that is less than £5. Ask them to take away the value of this item to see what change they would get from £5. Could they buy another item with the change? Keep going until they have 'spent' most of the £5. As a challenge, see if they can 'spend' exactly £5 when buying more than one item.

2.7 Subtraction with 4-digit numbers (2)

Subtract numbers with up to 4-digits

1. Use the column method to calculate.

   (a) 6001 – 4050 = ☐

   (b) 7934 – 4825 = ☐

   (c) 8102 – 7999 = ☐

   (d) 6025 – 998 = ☐

2. Fill in the blanks to make these calculations correct. Show any digits that should be under the bottom line.

   (a)

   ```
       6   3   ☐   4
   -   ☐   4   7   3
   _____
           9   1   ☐
   ```

   (b)

   ```
       9   9   9   9
   -   ☐   ☐   ☐   ☐
   _____
           1   2   3   4
   ```

   (c)

   ```
       ☐   ☐   ☐   ☐
   -   3   2   5   4
   _____
       3   8   8   9
   ```

   (d)

   ```
       9   0   0   0
   -   ☐   0   ☐   3
   _____
       3   ☐   1   ☐
   ```

👪 Time to explain

Look at the problem below with your child. Decide together whether it has been correctly calculated. Let your child explain mathematically their thoughts and correct it if necessary.

```
        5   ⁴5̸   ¹4   4
    -   4   4    5    5
    _____
        1   0    9    9
```

2.8 Estimating and checking answers using inverse operations

Estimate and check answers to a calculation

1. Use each given fact below to identify the other addition and subtraction facts.

   (a) $1106 - 225 = 881$

   $\boxed{} + 881 = \boxed{}$

   $1106 - \boxed{} = \boxed{}$

   $\boxed{} + 225 = \boxed{}$

   (b) $5237 - 1555 = 3682$

   $3682 + \boxed{} = \boxed{}$

   $\boxed{} + \boxed{} = 5237$

   $1555 = \boxed{} - \boxed{}$

   (c) $5726 - 1107 = \boxed{}$

   $1107 = \boxed{} \bigcirc \boxed{}$

   $\boxed{} = 1107 \bigcirc \boxed{}$

   $5726 = \boxed{} \bigcirc \boxed{}$

2. Fill in the boxes, using your understanding of addition and subtraction.

(a) $3546 + \boxed{} = 4634$

(b) $\boxed{} - 1074 = 3372$

(c) $4250 = \boxed{} + 3960$

(d) $4612 - \boxed{} = 3550$

(e) $6499 = 1850 + \boxed{}$

(f) $4256 + \boxed{} = 5247$

(g) $9632 + 1232 - \boxed{} = 3597$

(h) $8812 - 4565 - \boxed{} = 3681$

3. Estimate to the nearest 100 and then calculate.

(a) $4067 + 1999$

Estimate: $\boxed{}$

Calculate:

(b) $6888 - 2754$

Estimate: $\boxed{}$

Calculate:

👪Application

Solve this problem together. Which method is most efficient?

Paul runs a 1900-metre race at school. He completes the race in 4 minutes. He runs 401 metres in the first 2 minutes and 264 in the third minute. How far did he run in the last minute?

3.1 Multiplying whole tens by a 2-digit number

Multiply numbers by multiples of 10

1. Calculate with reasoning.

(a) 37 × 20 = ☐

(b) 40 × 18 = ☐

(c) 400 × 28 = ☐

(d) 90 × 50 = ☐

(e) 600 × 55 = ☐

(f) 60 × 550 = ☐

2. Identify the links between each calculation and write the answers.

(a) 15 × 2 = ☐

(b) 60 × 5 = ☐

(c) 72 × 3 = ☐

150 × 2 = ☐

6 × 500 = ☐

720 × 3 = ☐

15 × 20 = ☐

60 × 50 = ☐

72 × 30 = ☐

150 × 200 = ☐

600 × 5 = ☐

72 × 300 = ☐

👪 How does it work?

Ask your child to explain mathematically why the following is true or false. Can they make up another statement that would work in a similar way?

560 × 12 = 56 × 120

©HarperCollinsPublishers 2019

3.2 Multiplying a 2-digit number by a 2-digit number (1)

Use different methods to multiply 2-digit numbers together

1. Calculate these using a method of your choice.

   (a) 32 × 22 = ⬚

   (b) 81 × 93 = ⬚

   (c) 45 × 54 = ⬚

   (d) 73 × 27 = ⬚

   (e) 77 × 28 = ⬚

   (f) 67 × 45 = ⬚

2. Write the number sentences and calculate.

   (a) What is the product of 27 and 62?

   _____

   (b) Which is more: 39 × 31 or 35 × 35?

   _____

   (c) How much more is 600 than 18 times 24?

   _____

   (d) If the product is 1755 and one factor is 45, what is the other factor?

   _____

👨‍👦 Breaking it down

Talk with your child about why 36 × 14 is the same as:

   30 × 14 and 6 × 14 added together

   4 lots of 9 × 14

   36 × 10 and 36 × 4 added together

Ask them how else they could find the product of 36 × 14.

3.3 Multiplying a 2-digit number by a 2-digit number (2)

Use formal written methods to multiply 2-digit numbers together

1. Use the column method to calculate the following.

   (a) 15 × 39 = [____]

   (b) 63 × 17 = [____]

   (c) 74 × 13 = [____]

   (d) 47 × 32 = [____]

2. Find the error and make the corrections

   (a)
   ```
         5  7
   ×     3  6
   ─────────────
      7  0  2
      1  7  7
   ─────────────
      8  7  3
   ```

   (b)
   ```
            1  4
   ×        1  8
   ─────────────
         8  3  2
      1  4  0  0
   ─────────────
      2  2  3  2
   ```

3. Explain what the errors were in Question 2 and why they were wrong.

 Different calculations

Make your own set of 1–9 number cards with your child. Ask your child to randomly select four cards. Once they have these four single digits, they should use them to make a 2-digit by 2-digit multiplication. Challenge them to use the four digits to make as many different calculations as they can. What is the highest product they can achieve?

3.4 Multiplying a 3-digit number by a 2-digit number (1)

Use different methods to multiply 3-digit numbers by 2-digit numbers

1. Use the column method to calculate.

   (a) 111 × 32 =

   (b) 101 × 42 =

   (c) The product of 412 and 21.

   (d) The sum of 83 four hundred and fives.

   (e) 56 × 234 =

   (f) Multiplicand is 278. Multiplier is 49.

2. Draw lines to match the answer with the number sentence.

   281 × 24 =   12 096

   379 × 18 =   81 709

   672 × 18 =   6744

   505 × 26 =   13 130

   809 × 101 =   6822

 Many options

Challenge your child to find out how many different pairs of factors they can think of that give a product of 8000.

3.5 Multiplying a 3-digit number by a 2-digit number (2)

Use formal written methods to multiply 3-digit numbers by 2-digit numbers

1. Calculate mentally and write the answers.

(a) 32 × 4 = ☐

32 × 40 = ☐

32 × 140 = ☐

320 × 14 = ☐

(b) 46 × 5 = ☐

460 × 5 = ☐

46 × 50 = ☐

46 × 15 = ☐

(c) 189 × 75 = ☐

189 × 750 = ☐

1890 × 75 = ☐

2. Use column method to calculate.

(a) When 372 is added twenty-six times, the sum is?

(b) What is the product of 89 and 251?

(c) A factor is 99 and the product is 1584, but what is the other factor?

(d) 80 × 79 = 6320. What is the easy way to find 79 × 79?

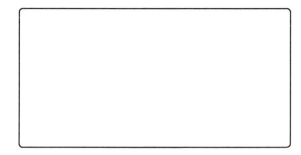

 Calculating smartly

Look at Question 2d above. Together, make up a similar question.

3.6 Dividing 2-digit or 3-digit numbers by tens

Use different methods to divide 3-digit numbers by multiples of ten

1. Use the column method to calculate.
   (a) 125 ÷ 30 =   (b) 472 ÷ 80 =   (c) 990 ÷ 10 =

   (d) 909 ÷ 60 =   (e) 824 ÷ 70 =   (f) 1234 ÷ 40 =

2. Complete the table to help the florist create bunches of flowers.

Type	Number ordered	Number in a bunch	Number of bunches (and any remainder)
lily	6528	30	
tulip	8372	50	
sunflower		20	52 r18
rose	7104	40	

 Pattern in number

Give your child the number 6482. Ask them to divide it by every multiple of ten between 0 and 99. Do they notice any pattern?

3.7 Practice and exercise

Use different methods to multiply and divide 3-digit numbers and 2-digit numbers

1. Draw lines to match related multiplication and division facts

   576 ÷ 16 =   12 × 143 =

   6552 ÷ 117 =   37 × 209 =

   1716 ÷ 13 =   36 × 16 =

   7733 ÷ 37 =   56 × 117 =

2. Using the relationship between multiplication and division, find the missing numbers in these calculations. Write down your calculations underneath.

   (a) ☐ ÷ 54 = 18

   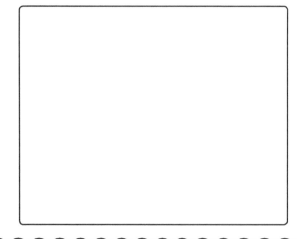

   (b) 66 × ☐ = 1122

   (c) ☐ × 14 = 406

   (d) 949 ÷ ☐ = 73

 Explaining mathematically

Ask your child to explain the relationships between the multiplication and division facts that have been paired in Question 1 above. Discuss, mathematically, how the facts are related.

Recognise and use hundredths

1. Draw lines to match the equivalent fractions.

$\dfrac{20}{100}$	$\dfrac{4}{5}$
$\dfrac{50}{100}$	$\dfrac{2}{4}$
$\dfrac{75}{100}$	$\dfrac{1}{10}$
$\dfrac{80}{100}$	$\dfrac{2}{10}$
$\dfrac{10}{100}$	$\dfrac{3}{4}$

2. Place these fractions on the blank number line below.

0

$\dfrac{1}{10}$ $\dfrac{2}{4}$ $\dfrac{89}{100}$ $\dfrac{4}{5}$ $\dfrac{70}{100}$ $\dfrac{7}{100}$ $\dfrac{2}{3}$

👪 Counting in hundredths

Work with your child to count in hundredths from any given starting point. Count forwards, backwards, in hundredths or in any other multiple of hundredths.

4.2 Addition and subtraction of fractions (1)

Add fractions with the same denominator

1. Calculate.

   (a) $\frac{1}{5} + \frac{3}{5} = \boxed{}$

   (b) $\frac{32}{50} + \frac{17}{50} = \boxed{}$

   (c) $\frac{18}{30} + \frac{12}{30} = \boxed{}$

   (d) $\frac{42}{100} + \frac{37}{100} = \boxed{}$

   (e) $\frac{250}{600} + \frac{99}{600} + \frac{101}{600} = \boxed{}$

   (f) $\frac{12}{40} + \frac{13}{40} + \frac{14}{40} = \boxed{}$

2. Calculate.

   (a) The sum of $\frac{16}{34}$ and $\frac{12}{34}$. $\boxed{}$

   (b) 9 lots of $\frac{1}{15}$ add 4 lots of $\frac{1}{15}$ equals $\boxed{}$.

   (c) $\frac{14}{42} + \boxed{} = \frac{37}{42}$

   (d) How many times would I need to repeatedly add $\frac{2}{5}$ to get to a whole? $\boxed{}$

   (e) $\boxed{} + \frac{12}{25} = 1$

   (f) $\frac{19}{35} + \boxed{} = \frac{34}{35}$

Difference between fractions

Ask your child: *After $\frac{5}{17}$ is taken away from a number, $\frac{8}{17}$ is left. What was the starting number?* Your child should draw a diagram to show this.

4.3 Addition and subtraction of fractions (2)

Add fractions with the same denominator

1. Use the diagrams given to complete the calculations.

(a) $\frac{4}{7} + \frac{2}{7} = \boxed{}$

(b) $\frac{6}{14} + \frac{5}{14} = \boxed{}$

(c) $\frac{15}{20} - \frac{9}{20} = \boxed{}$

(d) $\frac{13}{15} - \boxed{} = \frac{9}{15}$

2. Calculate.

(a) The minuend is $\frac{11}{25}$ and the subtrahend is $\frac{5}{25}$.
What is the difference?

(b) What is the difference when subtracting $\frac{80}{100}$ from 1?

(c) $1 - \frac{12}{25} - \frac{10}{25} =$

(d) $1 - \frac{1}{2} - \frac{5}{30} =$

(e) 5 lots of $\frac{1}{8}$ subtract 4 lots of $\frac{1}{8}$.

(f) $\boxed{} - \frac{15}{25} = \frac{8}{25}$

👪 Saving chocolate!

Explain to your child that Sam likes to make his chocolate last as long as possible. Each day he eats half of what he has. Ask your child to create a chart to show how much of the original bar he eats each day.

Day	1	2	3	4	5	6	7	8	...
He eats ...									

4.4 Fun with exploration – 'fraction wall'

Compare, add and subtract fractions

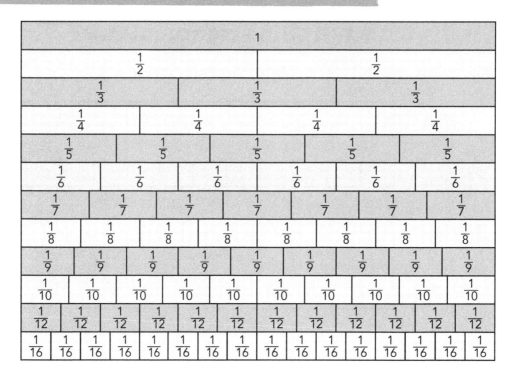

1. Identify fractions that are the same as those given.

   (a) $\frac{4}{12}$ is equal to …

   (b) $\frac{6}{9}$ is equal to …

   (c) $\frac{12}{16}$ is equal to …

   (d) $\frac{4}{7}$ is equal to …

   (e) $\frac{8}{16}$ is not equal to …

   (f) $\frac{7}{10}$ is smaller than …

2. Use <, > or = to show the relationship between these fractions.

   (a) $\frac{5}{7}$ ☐ $\frac{5}{9}$

   (b) $\frac{14}{16}$ ☐ $\frac{10}{12}$

   (c) $\frac{1}{2}$ ☐ $\frac{3}{6}$

   (d) $\frac{5}{12}$ ☐ $\frac{5}{9}$ ☐ $\frac{9}{16}$

   (e) $\frac{1}{3}$ ☐ $\frac{1}{5}$ ☐ 1

   (f) $\frac{3}{4}$ ☐ $\frac{6}{8}$ ☐ $\frac{9}{12}$

 Fraction wall

Help your child to create their own fraction wall. As they draw it, ask them to talk about the relationship between different sfractions using the language of numerator and denominator to describe each fraction.

5.1 Multiplication and multiplication tables

Recall multiplication facts and solve multiplication problems

1. Write these repeated addition number sentences as multiplications.

   (a) $9 + 9 + 9 + 9 + 9 + 9 + 9 =$ ☐ $\times$ ☐ $=$ ☐

   (b) $7 + 7 + 7 + 7 + 7 + 7 + 7 + 7 + 7 =$ ☐ $\times$ ☐ $=$ ☐

   (c) $5 + 5 + 5 + 5 + 5 + 5 =$ ☐ $\times$ ☐ $=$ ☐

   (d) _____ $= 9 \times 13 =$ ☐

   (e) _____ $= 6 \times 15 =$ ☐

   (f) $8 + 8 + 8 + 8 + 8 + 8 + 8 + 8 =$ ☐ $\times$ ☐ $=$ ☐

2. Fill in the answers and write two multiplication sentences and two division sentences for each of the following (the first one is done for you):

   (a) 3 times 9 is ☐

   $3 \times 9 = 27$ $27 \div 9 = 3$
   $9 \times 3 = 27$ $27 \div 3 = 9$

   (b) 8 times 5 is ☐

   _____   _____

   _____   _____

   (c) 7 times 6 is ☐

   _____   _____

   _____   _____

3. Circle the numbers below that appear in the 2, 4 and 6 multiplication patterns. The first number is 12.

   2, 4, 6, 8, 10, 12, 14, 16, 18, 20, 22, 24, 26, 28, 30, 32, 34, 36, 38, 40

4. For all the numbers you have circled in Question 3, write the multiplication and division number sentences for each of the 2, 4 and 6 times tables.

For example, $4 \times 3 = 12$; $12 \div 4 = 3$.

Multiplication square

×	1	2	3	4	5	6	7	8	9	10	11	12
1												
2												
3												
4												
5												
6												
7												
8												
9												
10												
11												
12												

Time your child as they complete the multiplication square. How quickly can they complete all of the facts? Ask them to identify the facts that they are less sure about and practise these on their own, then time your child again. Have they improved?

5.2 Relationship between addition and subtraction

Use the inverse relationship between addition and subtraction

1. Write the number sentence and complete.

   (a) The difference between two numbers is 198. What could the minuend and subtrahend be? Can you find five possibilities?

   (b) The minuend is 999 and the difference is 332. What is the subtrahend?

   (c) The sum of two addends is 350. One addend is > 150 and the other is < 100. Identify four different possible addend combinations.

   (d) The subtrahend is 408 and the difference is 289. What is the minuend?

(e) The sum of three addends is 1069. If one of the addends is 512 and another 298, what is the third?

(f) Addend + addend = sum. What other generalisations do you know about addition and subtraction?

2. True or false? Put a ✓ for true and a ✗ for false in each box.

(a) If I add three 3-digit numbers, the sum must be greater than 1000. ☐

(b) In an addition problem, if the sum is an even number the addends must both be even. ☐

(c) In a subtraction, if the minuend is odd and the subtrahend is even, the difference is always even. ☐

(d) $309\,684 - 12\,574 = 309\,684 + 12\,574$ ☐

(e) $A + B + C = D$ and $D - A - B = C$ ☐

(f) $9999 - 333 = 10\,000 - 334$ ☐

👪 Relationships

Together, create a line model to show this addition number sentence: $53 + 48 + 31 =$ ☐

After your child has created a line model, ask them to write down all of the related number facts that they know from the information given.

5.3 The relationship between multiplication and division

Use the inverse relationship between multiplication and division

1. Use the relationship between multiplication and division to complete these problems.

   (a) $70 \times \boxed{} = 1120$

   (b) $\boxed{} \times 44 = 1672$

   (c) $1131 \div \boxed{} = 87$

   (d) $9936 \div \boxed{} = 108$

   (e) $\boxed{} \div 46 = 112$

   (f) $\boxed{} \times 74 = 2146$

2. Draw lines to match related facts, and fill in the blanks.

   $123 \times \boxed{} = 5535$

   $\boxed{} \times 753 = 11\,295$

   $\boxed{} \times 15 = 11\,295$

   $\boxed{} \div 698 = 14$

   $698 \times 14 = \boxed{}$

   $\boxed{} = 7200 \div 18$

   $109 \times \boxed{} = 3270$

   $\boxed{} \div 123 = 45$

   $\boxed{} \times 18 = 7200$

   $3270 \div \boxed{} = 109$

What is the link?

Have a mathematical discussion with your child. Discuss the link between multiplication and division. Help your child to create a poster that explains this link. They can use examples to help clearly show their explanation.

Use written methods to multiply by 2-digit numbers

1. Without calculating, complete the following.

(a) $49 \times 112 = 56 \times$ [ ]

(b) [ ] $\times 168 = 90 \times 84$

(c) $242 \times 32 = 16 \times$ [ ]

(d) $290 \times$ [ ] $= 145 \times 34$

2. Fill in the spaces.

(a) The product of 267×23 is a [ ]-digit number.

(b) The digits in the product of 462×14 are all _____ (odd/even).

(c) The product of 234×19 rounded to the nearest hundred is [ ].

(d) If one factor is an odd number, the product
   will _____ (sometimes, always, never) be odd.

(e) 243 must be multiplied by [ ] to make the answer
   as close as possible to 10 000.

~~~~~~

Household bills

Help your child to find out the amount that you spend as a family on individual bills a week/month and then ask them to calculate the amount that is spent across the month/year.

Which bills does the family spend the most money on?

5.5 Practice with fractions

Calculate fractions of amounts and add and subtract fractions

1. What fraction of each shape is shaded?

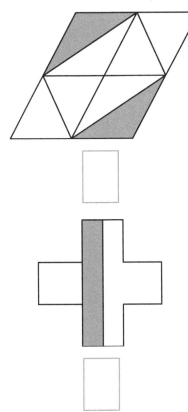

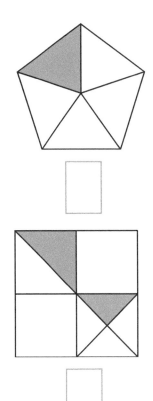

2. Put these fractions in order on the number line below.

$$\frac{1}{4}, \frac{20}{100}, \frac{6}{8}, \frac{3}{5}, \frac{9}{10}$$

0

👪 Fraction combinations

Using the image below, help your child to identify the fraction of the shape that is shaded.

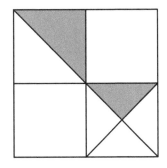

1. These are sections of a I–C (1–100) Roman numeral grid.
   Complete the sections.

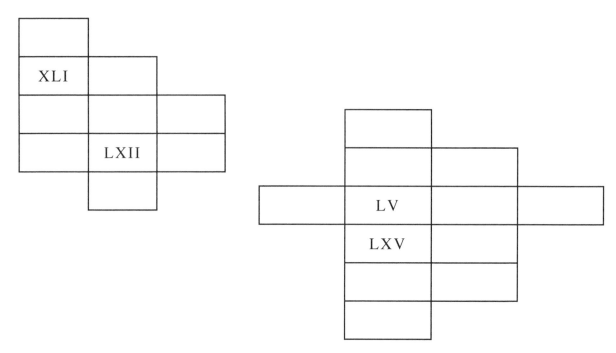

2. Write these numbers as Roman numerals.

(a) 55 = _____

(b) 90 = _____

(c) 8 = _____

(d) 18 = _____

(e) 28 = _____

(f) 33 = _____

(g) 101 = _____

(h) 14 = _____

(i) 114 = _____

(j) 200 = _____

👪 Roman numerals today

Discuss with your child how and where Roman numerals are still used today. If you have access to a camera, let your child take photographs of the Roman numerals they find. They can then sort the photos in order, from smallest to greatest.

6.1 Decimals in life

Recognise decimal numbers in the context of money and measures

1. Fill in the spaces.

(a) We say £15.30 as _____ .

(b) We say 3.4 kg as _____ .

(c) _____ is read as fifty-six pounds and eighty-three pence.

(d) _____ is read as thirty-three point six litres.

(e) My height is 1.79 metres which is read as _____ .

(f) The price of petrol is _____ which is read as one hundred and twenty-three point nine pence.

Decimals around us

With your child, look at food in your kitchen cupboards. List the weight or volume of 10 different packets, tins, bottles, or jars that you find.

Food	Type of container – packet/tin/bottle/jar	Weight or volume

6.2 Understanding decimals (1)

Recognise and write decimal equivalents to tenths, hundredths, thousandths, halves and quarters

1. Use each grid to show the hundredths fraction and write the decimal equivalent. The first one is done for you.

(a) $\frac{10}{100} = 0.10 = 0.1$

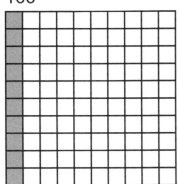

(b) $\frac{1}{4} =$ _____

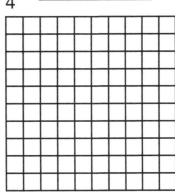

(c) $\frac{1}{2} =$ _____

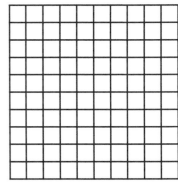

(d) $\frac{3}{4} =$ _____

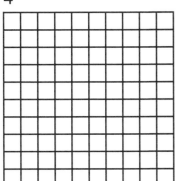

(e) $0.3 =$ _____

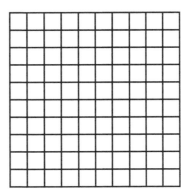

(f) $\frac{4}{10} =$ _____

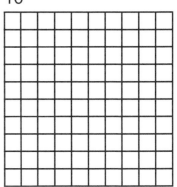

2. Write the following decimals as fractions.

(a) $0.7 =$ ☐   (b) $0.07 =$ ☐   (c) $0.16 =$ ☐   (d) $0.32 =$ ☐

 Decimal wall

Place 10 crayons or felt pens on a table for your child to look at – use four different colours. Ask your child to complete the table below.

Colour	Fraction	Decimal

6.3 Understanding decimals (2)

Recognise and write decimal equivalents to tenths, hundredths and thousandths

1. Count in decimals and complete the patterns.

   (a) Count in 0.1s.

   0.5, 0.6, ⬚, ⬚, ⬚, ⬚, 1.1, ⬚, ⬚, ⬚

   (b) Count in 0.01s.

   0.93, 0.94, ⬚, ⬚, ⬚, ⬚, ⬚, ⬚, ⬚

   (c) Count in 0.2s.

   2.2, ⬚, ⬚, ⬚, ⬚, 3.2, ⬚, ⬚, ⬚

   (d) Count in 0.05s.

   ⬚, ⬚, ⬚, ⬚, 0.35, ⬚, ⬚, ⬚, ⬚

2. Fill in the spaces to complete the statements.

   (a) The fraction $\frac{25}{100}$ can be written as a decimal as _____ or _____.

   (b) A number with 42 tenths is _____ (less/more) than 4.

   (c) The number _____ consists of 8 tenths and 11 thousandths.

   (d) The decimal _____ when written as a fraction is $\frac{20}{100}$.

   (e) There are four _____, six _____,

   five _____, three _____ and

   eight _____ in the number 46.538.

👪 Whole numbers

All whole numbers can be partitioned into decimal parts. Ask your child to take each whole number between 1 and 10 on its own and generate a number sentence using any of the operations (+, −, ×, ÷) with the whole number as the total. For example, 4 × 0.25 = 1. They should find at least four sentences for each number up to 10.

6.4 Understanding decimals (3)

Recognise and write decimal equivalents to tenths, hundredths and thousandths

1. Fill in the spaces to complete.

(a) 0.876 = ☐ × 0.1 + ☐ × 0.01 + ☐ × 0.001

(b) 0.054 = ☐ × 1 + ☐ × 0.1 + ☐ × 0.01 + ☐ × 0.001

(c) 2.803 = ☐ × 1 + ☐ × 0.1 + ☐ × 0.01 + ☐ × 0.001

(d) 42.51 = ☐ × 10 + ☐ × 1 + ☐ × 0.1 + ☐ × 0.01

(e) ☐ = 6 × 10 + 5 × 1 + 7 × 0.1 + 3 × 0.001

2. Use the symbols >, < and = to express the relationship between these.

(a) pure decimals ☐ mixed decimals

(b) £14.80 ☐ £14.08

(c) $2\frac{3}{5}$ ☐ 2.6

(d) 654 ☐ 65.4

(e) $\frac{150}{200}$ ☐ 0.386

👪 Pure or mixed?

While you are out with your child, search for examples of mixed and pure decimal numbers. Ask your child to make a tally to show the frequency of each of these numbers seen in one day.

6.5 Understanding decimals (4)

Recognise and write decimal equivalents to tenths, hundredths and thousandths

1. Complete the Carroll diagram below with these decimal numbers.

0.78, 32.87, 328.7, 0.5, 0.123, 5.405, 6.50, 0.004, 0.710, 1073.7

	One decimal place	Two decimal places	Three decimal places
Mixed decimal			
Pure decimal			

2. Convert these proper and improper fractions into decimals.

(a) $1\frac{5}{20}$ = ☐

(b) $\frac{25}{20}$ = ☐

(c) $\frac{32}{10}$ = ☐

(d) $4\frac{13}{25}$ = ☐

(e) $\frac{190}{100}$ = ☐

(f) $\frac{1900}{1000}$ = ☐

👪 Key concept

Ask your child how well they think they understand proper and improper fractions. Can they tell you that proper fractions are those in which the numerator is smaller, or equal to, the denominator (and therefore it shows that it is part of a whole), whereas improper fractions are those in which the numerator is bigger than the denominator and indicates that it is more than one whole?

6.6 Understanding decimals (5)

Use decimals to convert between different units of measure

1. Draw the given objects against the ruler to the correct size.

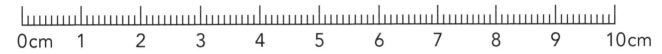

0 cm   1   2   3   4   5   6   7   8   9   10 cm

(a) length of a glue stick = 73 mm

(b) length of a rubber = 4.5 cm

(c) length of a pencil = $\frac{90}{1000}$ m

(d) diameter of a £1 coin = 0.02 m

2. Write each measurement in the correct group below.

$\frac{98}{100}$ m, 3.8 cm, $\frac{1000}{10}$ cm, 0.67 m, 6.07 m, 6.7 cm, 0.001 km

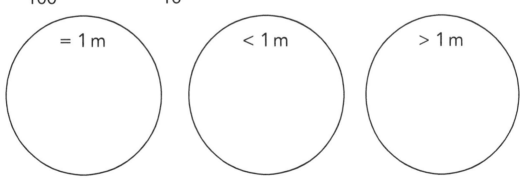

= 1 m   < 1 m   > 1 m

👪 Measurements

Using a ruler or metre tape, help your child to measure objects around the house that are less than a metre. Ask them to record the measurements as a decimal number. Set each other challenges to find something smaller/bigger than a given measurement and then measure accurately to see if your prediction is correct.

6.7 Understanding decimals (6)

Recognise and write decimal equivalents of tenths, hundredths and thousandths

1. Multiple choice questions. For each question, choose the correct answer and write the letter in the box.

   (a) $\frac{14}{10}$ as a decimal is: ☐

       A. 0.14   B. 14   C. 0.014   D. 1.4

   (b) 105×0.1 is: ☐

       A. 10.5   B. 100.5   C. 1.05   D. 105

   (c) 0.4 as a fraction is: ☐

       A. $\frac{4}{100}$ B. $\frac{4000}{1000}$ C. $\frac{40}{10}$ D. $\frac{40}{100}$

   (d) The largest decimal number created with the digits 5, 6 and 7 is: ☐

       A. 7.65   B. 76.5   C. 0.765   D. 765

2. Guess the number using the clues below. It is a mixed number with 3 decimal places.

   The whole number is 15 less than $\frac{4}{10}$ of 100.

   The tenths digit is the sum of both the digits in the whole number.

   The hundredths digit is an even number greater than the tenths digit.

   The thousandths digit is half the hundredths digit.

   The number is _____.

~~~~~~

👫 Decimal numbers to 3 decimal places

Play this game with your child. Take it in turns to roll a dice. The number that the dice lands on becomes your number (for example, if you rolled 5 on your first go, your number is 5). Keep rolling to generate digits for a decimal number to 3 decimal places. Swap the digits around to make the biggest possible number with the digits you have. Who can create the largest number?

6.8 Comparing decimals (1)

Compare and order decimals

1. Use >, < and = to compare the following decimal numbers.

   (a) 0.99 ☐ 0.111

   (b) $\frac{105}{100}$ ☐ 1.05

   (c) 0.123 ☐ 0.12

   (d) 7.4 ☐ 7.44

   (e) $\frac{43}{10}$ ☐ 4.03

   (f) 9.99 ☐ 10

2. Put these decimal numbers in order, starting with the smallest.

   (a) 8.5, 8.05, 8.51, 8.15

   ☐

   (b) 0.804, 0.084, 0.408, 0.048

   ☐

   (c) 6.5, 6.48, 6.84, 6.54

   ☐

   (d) 0.727, 0.707, 0.72, 0.7

   ☐

   (e) 8.88, 8.8, 8.08, 8.808

   ☐

   (f) 3.612, 36.12, 3.12, 3.62

   ☐

👪 Heights

Help your child to measure the heights of members of your family or friends. Ask them to record these measurements as decimal numbers using metres and centimetres. Once they have five measurements, challenge them to order them from smallest to largest.

6.9 Comparing decimals (2)

Compare, order and round decimal numbers

1. Fill in the boxes.

(a) Rounding 13.4 to the nearest whole number, the result is ☐.

(b) Rounding 0.12 to the nearest whole number, the result is ☐.

(c) Rounding 1.04 to the nearest whole number, the result is ☐.

(d) Rounding 99.06 to the nearest whole number, the result is ☐.

(e) Rounding 50.7 to the nearest whole number, the result is ☐.

(f) Rounding 18.0 to the nearest whole number, the result is ☐.

2. Place these decimal numbers into the correct group.

3.67, 5.09, 5.1, 603.8, 123.4, 12.12, 12.52, 99.9, 1005.4

Round up to the
next whole number

Round down to the
previous whole number

👪 Key concept

It is important to remember the generalised rule of rounding. If the number in a column is 5 or above, we round up but if it is 4 or below, we round down. When rounding decimal numbers to a whole number, we first look at the digit in the tenths place. If it is not possible to round using the tenths digit, then we look at the hundredths digit and follow the rule as above.

Check your child's understanding by giving them some decimal numbers and ask them which ones they would round down and which they would round up.

6.10 Properties of decimals

Identify properties of decimals, including the value of any zeros

1. Use what you understand about decimals to simplify the following.

   (a) 500.010 = ☐   (b) 41.100 = ☐

   (c) 450.10 = ☐   (d) 1000.120 = ☐

   (e) 80.900 = ☐   (f) 15.707 = ☐

   (g) 1000.00 = ☐   (h) 17.100 = ☐

2. Compare the numbers by writing >, < or = in the boxes.

   (a) 5.50 ☐ 5.5   (b) 100.405 ☐ 100.450

   (c) 8.35 ☐ 8.350   (d) 15.90 ☐ 15.92

   (e) $\frac{25}{100}$ ☐ 0.21   (f) $\frac{25}{10}$ ☐ 2.05

   (g) 145.26 ☐ 145.206   (h) 25.50 ☐ 25.05

👪 Learning summary

Discuss with your child what they have learned through this decimals sequence of lessons. What have been the key pieces of learning for your child and what has been challenging for them to understand? What aspects of working with decimals will they be continuing to practise and which are they feeling really confident with?

Interpret information presented in line graphs and time graphs

1. Look at the line graph below and answer the questions.

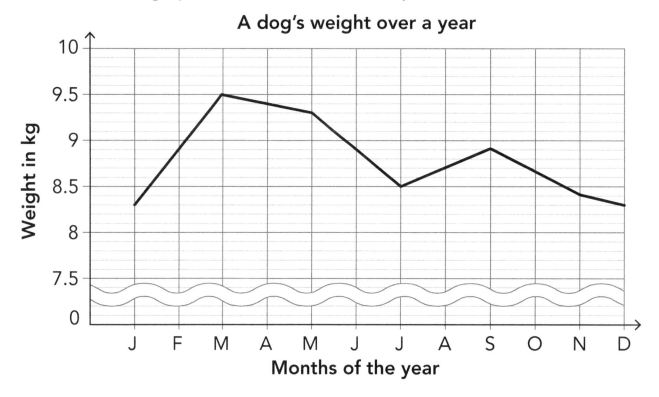

A dog's weight over a year

(a) Over which months did the dog lose the most weight?

_____

(b) Over which month did the dog put weight back on?

_____

(c) In which month was the dog at its lightest?

_____

(d) How long did it take for the dog to lose 1.2 kg from its heaviest?

_____

(e) How much weight did the dog lose between months 3 and 7?

_____

👪 **Recording temperature**

Together with your child, measure the temperature outside over the course of a day. Record your measurements on a line graph of your creation. Help your child to label the axes accurately and represent the information carefully.

7.2 Knowing line graphs (2)

Interpret information presented in line graphs and bar charts

1. Interpret the line graph and answer the questions below.

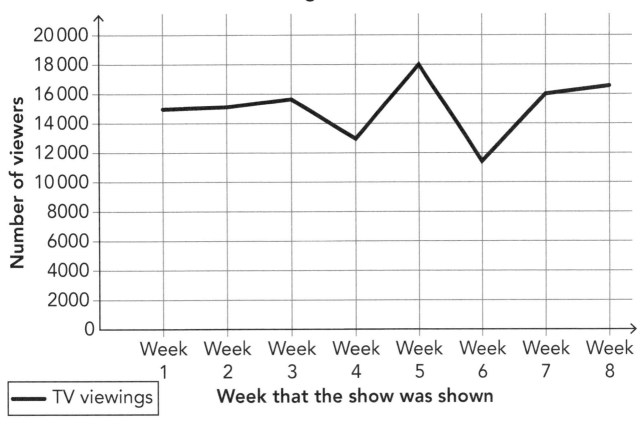

Number of viewers watching the same TV show over 8 weeks

(a) In which week was the show watched by the most people?

_____

(b) In which week was the show watched by the fewest people?

_____

(c) In which weeks did the show have more than 14 000 viewers?

_____

(d) In which weeks did the show have less that 16 000 viewers?

_____

👪 Discussion

Line graphs are a useful and important way to track changes over a period of time.
Ask your child if they can think of examples where line graphs are useful to help us
understand how something changes over time.

Interpret information presented in line graphs

1. Explore the line graph and complete the questions below.

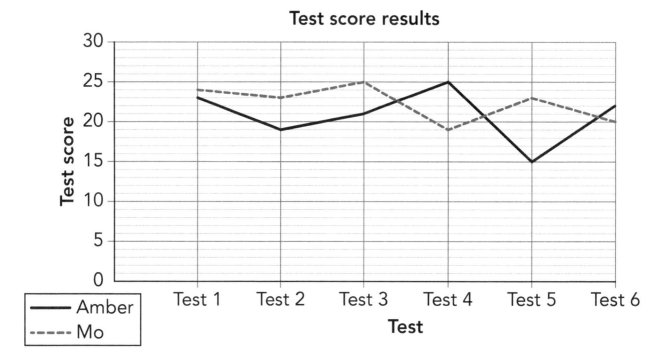

Test score results

(a) What was the sum of all test scores for:

Amber [   ]   Mo? [   ]

(b) In which tests did Amber get a better score than Mo?

_____

(c) In which test was there the biggest difference in scores?

_____

(d) Which tests have a difference of 4?

_____

(e) For which test(s) was there the best overall score?

_____

(f) This line graph remains untitled, what title would you give it?

_____

👪Comparing data

The line graph above includes two sets of data that can be compared. Discuss with your child when in everyday life we might want to use line graphs to compare different sets of data. Ask them to write some examples they can think of.

7.4 Constructing line graphs

Construct line graphs and interpret information from them

1. Use your understanding of line graphs to construct your own. Use the template provided, or draw your own. Decide what you would like to collect data about and record your findings here.

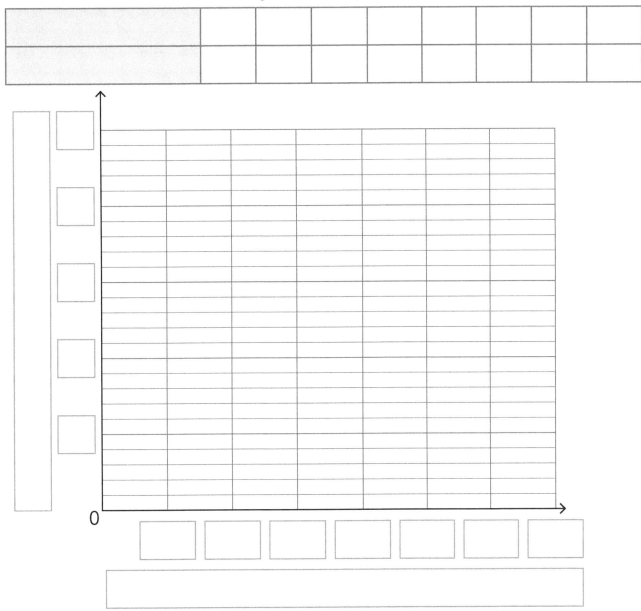

Interpreting your line graph

After your child has completed their line graph, explore it together and consider the following questions.

(a) Were there any surprises about your data?

(b) What is your data telling you?

(c) Which is the greatest/lowest value?

(d) When was there the greatest change in values?

(e) What title would you give to your line graph?

(f) If you repeated the collection of your data, would you get the same results?

8.1 Acute and obtuse angles

Identify acute and obtuse angles

1. Label the angles identified as either acute, obtuse or right.

(a) _____

(b) _____

(c) _____

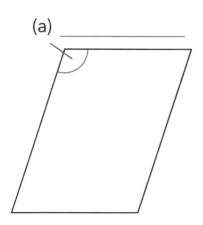
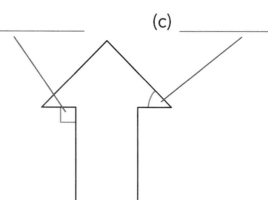

(e) _____

(d) _____

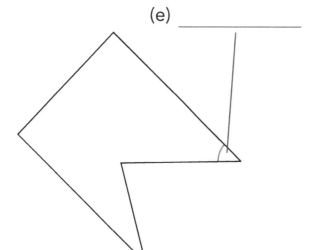
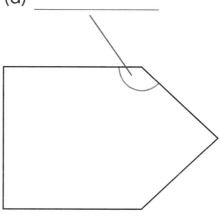

2. Draw lines to match the clues with the correct shape.

4 right angles   scalene triangle

1 right angle, 2 acute angles   right-angled triangle

3 acute angles   square

5 obtuse angles   pentagon

Angle hunt

Using an A4 piece of paper as an 'angle finder' (this is a good tool as the corners are right angles and can be compared with other angles that you find), encourage your child to explore objects and structures in your home. Ask them to make a list of at least 12 angles they find and say where they found them.

8.2 Triangles and quadrilaterals (1)

Understand the properties of triangles and quadrilaterals

1. Draw lines on each shape to complete the given instructions.

   (a) Inside shape A, create 1 quadrilateral and 2 triangles.

   (b) Inside shape B, create 3 triangles.

   (c) Inside shape C, create 1 quadrilateral and 3 triangles.

   (d) Inside shape D, create 1 quadrilateral and 3 triangles.

   (e) Inside shape E, create 1 quadrilateral and 4 triangles.

   (f) Inside shape F, create 3 quadrilaterals.

   (g) Inside shape G, create 2 quadrilaterals and 1 triangle.

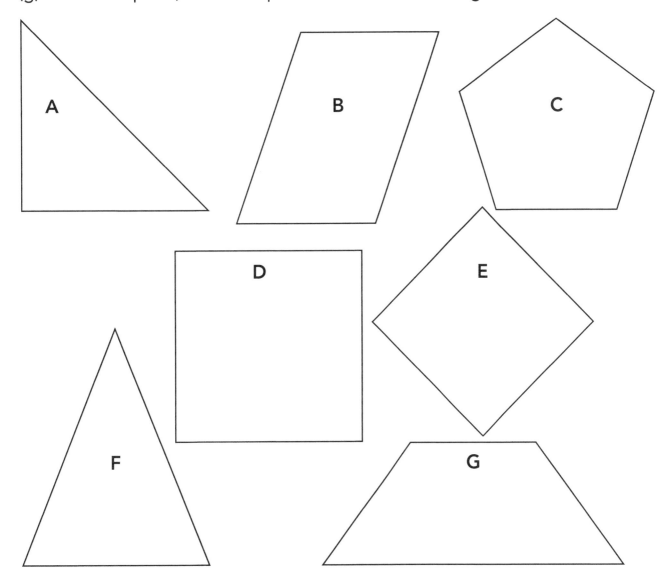

👪 Make a shape

Together, set each other challenges to draw lines within the above shapes. How many different examples can you identify together? For example, by drawing 4 lines, create 1 square and 3 triangles in shape F.

8.3 Triangles and quadrilaterals (2)

Understand the properties of triangles and quadrilaterals

1. Complete the shapes by drawing lines with a ruler. Use the clues to help you. Each shape will be either a quadrilateral or triangle.

   A: 2 lines. The shape created is a _____.

   B: 1 line. The shape created is a _____.

   C: 3 lines. The shape created is a _____.

   D: 2 lines to create a square. A square is a _____.

   E: 2 lines to create a parallelogram. A parallelogram is a _____.

   F: 2 lines to create a right-angled triangle.

   A right-angled triangle is a _____.

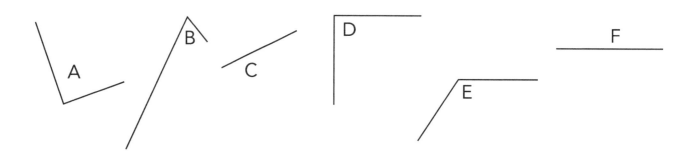

2. For each of the shapes above, record how many acute, obtuse and right angles you see.

   (a) Shape A = ☐ acute, ☐ obtuse, ☐ right angle.

   (b) Shape B = ☐ acute, ☐ obtuse, ☐ right angle.

   (c) Shape C = ☐ acute, ☐ obtuse, ☐ right angle.

   (d) Shape D = ☐ acute, ☐ obtuse, ☐ right angle.

   (e) Shape E = ☐ acute, ☐ obtuse, ☐ right angle.

   (f) Shape F = ☐ acute, ☐ obtuse, ☐ right angle.

👪 I challenge you ...

Each draw two shapes without letting the other person see what you have drawn. Give each other information about the number and type of angles in each of your shapes. You should draw your opponent's shapes. Can your child describe the shapes accurately enough for their opponent to be able to draw them correctly?

8.4 Classification of triangles (1)

Name and classify triangles in terms of angles

1. Group these triangles by writing the letter into the groups below.

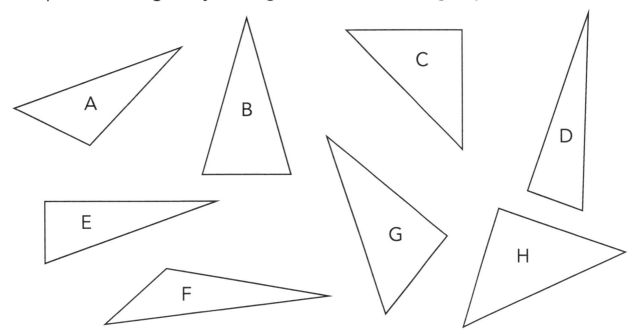

Right-angled triangles	Obtuse-angled triangles	Acute-angled triangles

2. Draw an acute-angled triangle, a right-angled triangle and an obtuse-angled triangle onto the grid below. The first line of each is drawn for you.

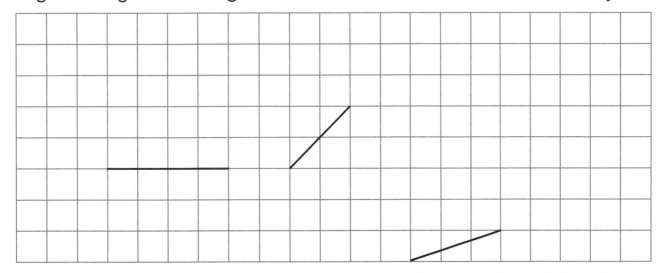

👪 Classification

Discuss the triangles from Question 1 and how your child grouped them. Can they accurately explain the difference between acute-angled, obtuse-angled and right-angled triangles?

8.5 Classification of triangles (2)

Name and classify triangles in terms of angles

1. Using the grid below, draw straight lines so that it contains:

   3 right-angled triangles
   2 obtuse-angled triangle
   2 acute-angled triangles

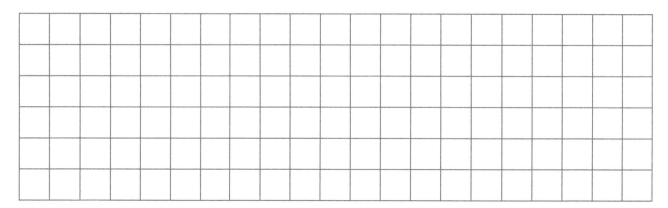

2. Draw lines in the squares to create triangles inside each square.

   Draw 1 line   Draw 2 lines   Draw 3 lines   Draw 4 lines   Draw 5 lines

   How many triangles did you create and what type of triangles were they?

	Number of triangles	Type of triangles
1 line		
2 lines		
3 lines		
4 lines		
5 lines		

👪 Triangle challenge

Ask your child to look again at Question 2. Talk about what they notice about the types of triangles created by drawing the lines through the square. Why do they think this is? Is there a pattern? What about obtuse-angled triangles – can they explain why there aren't any?

8.6 Line symmetry

Identify line symmetry in shapes and complete symmetrical figures

1. On the grid below, complete the shapes to make them symmetrical.

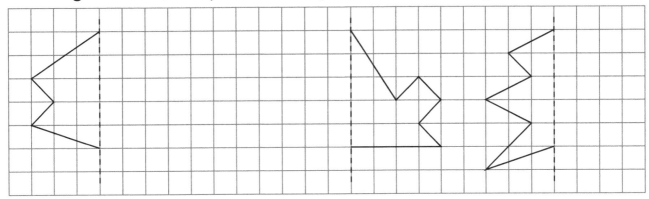

2. On the grid below, use the line of symmetry to reflect each shape. The first one has been completed for you.

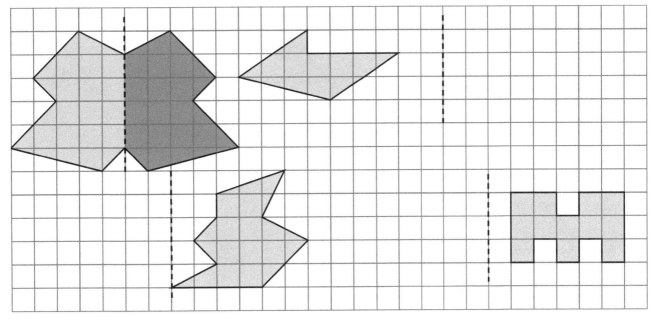

👫 Symmetrical or not?

Together, explore the world around you looking for things that are symmetrical and things that are not. Ask your child to make a list or draw the things that they find which are symmetrical.

Name and classify triangles in terms of sides

1. Write a definition for the following triangles:

   (a) scalene _____

   _____

   (b) isosceles _____

   _____

   (c) equilateral _____

   _____

2. Group these triangles according to their properties.

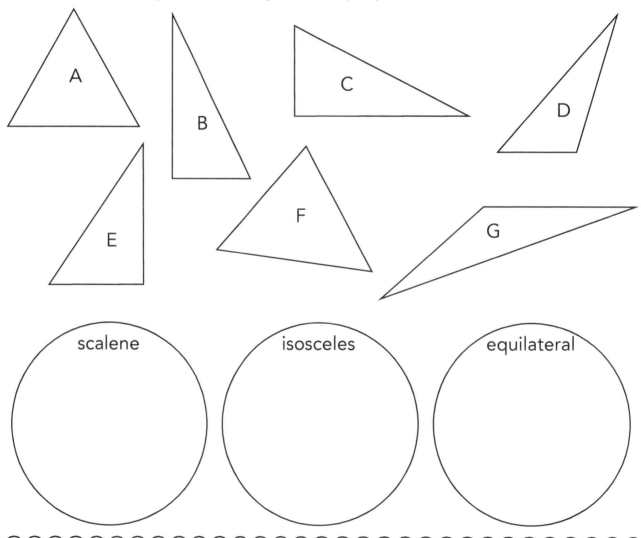

| scalene | isosceles | equilateral |

👪 Guess my triangle

Ask your child to explain what they know about the key properties of the triangles explored during this homework. Can they explain clearly what makes a triangle isosceles, or equilateral, or scalene?

8.8 Areas

Find the area of shapes by counting squares

1. Count the number of squares that each shape occupies on this grid.

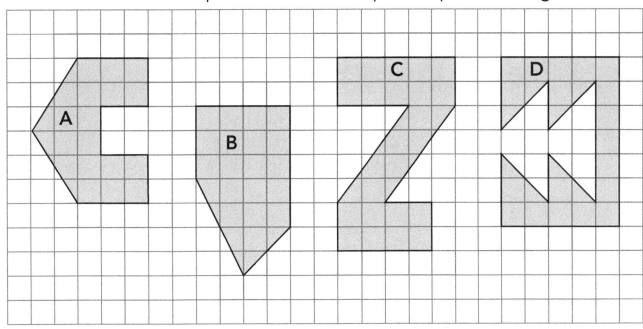

A = ☐ squares B = ☐ squares C = ☐ squares D = ☐ squares

2. Draw three shapes onto the grid below. They should occupy 9, 12 and 15 squares respectively. Challenge yourself.

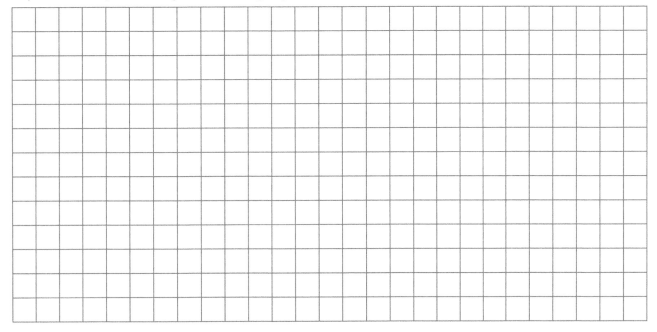

~~~~~~

Your house

On squared paper, help your child to create a floor plan of your house. Encourage them to try to make the rooms comparative in size. They should count the squares each room occupies and record it next to the rooms. Which room is the largest? Can they order the rooms based on their size?

©HarperCollinsPublishers 2019

8.9 Areas of rectangles and squares (1)

Calculate the areas of square and rectangles

1. Calculate the areas of these squares and rectangles.

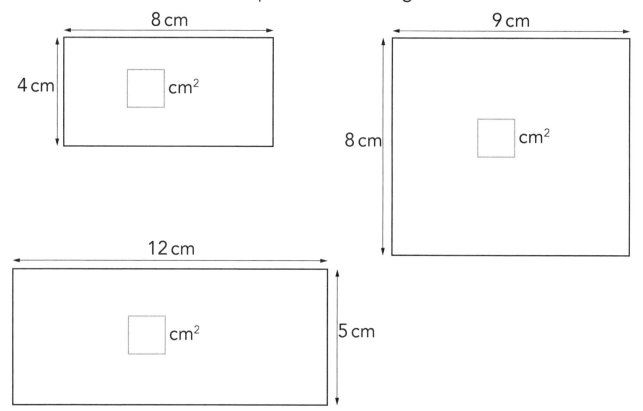

2. Calculate the area of these shapes using the clues given.

(a) The length of each side of a square is 9. What is its area?

Answer: _____

(b) The area of a rectangle is 48 cm. Its width is 8 cm. What is its length?

Answer: _____

(c) A book cover measures 14 cm from top to bottom. Its width is 7 cm less than this. What is the area of the book cover?

Answer: _____

(d) I put two squares together to make a rectangle. The area of one square is 36 cm². What is the length and width of the rectangle?

Answer: _____

Rectangle challenge

Solve this problem together. The area of a rectangle is 32 cm². Its length is double its width. What is its perimeter?

Answer: _____

Calculate the areas of squares and rectangles

1. Use the information below to fill in the spaces.

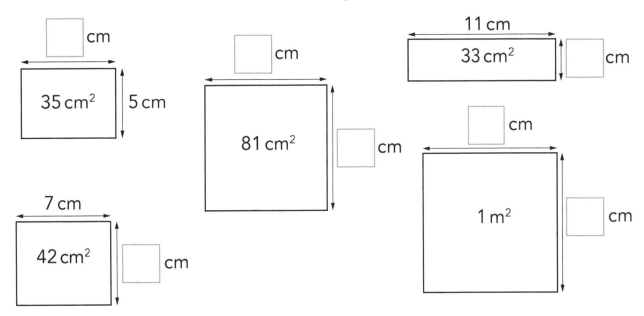

2. Draw a diagram to help you solve this problem.

   The length of a rectangle is 13 cm. The area is 52 cm². The rectangle is extended by the area 26 cm². What is the width of the rectangle before and after it is extended?

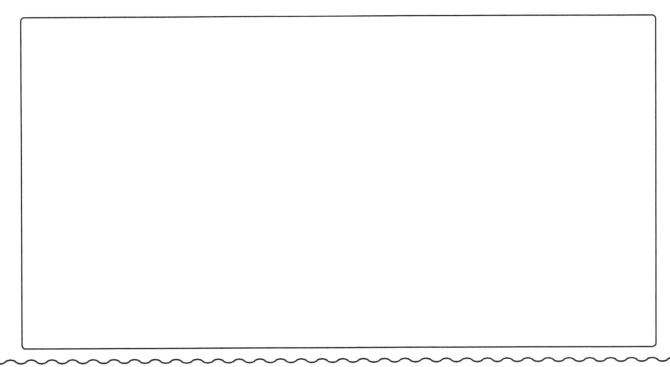

👪 Sometimes, always, never?

Is the following statement sometimes, always or never true? Ask your child to record any working out that they do to test their thinking.

The area of a square and a rectangle will always be equal.

8.11 Square metres

Calculate the areas of squares and rectangles using square metres

1. Calculate the area of the following diagrams.

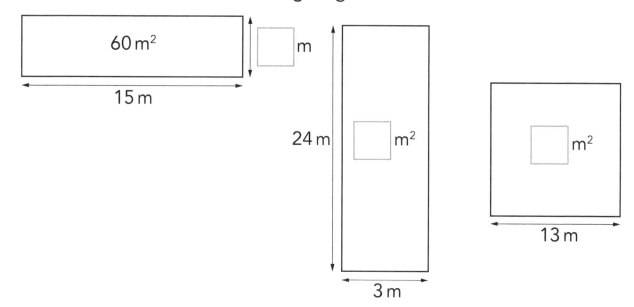

2. Calculate the area of the shaded part of each diagram.

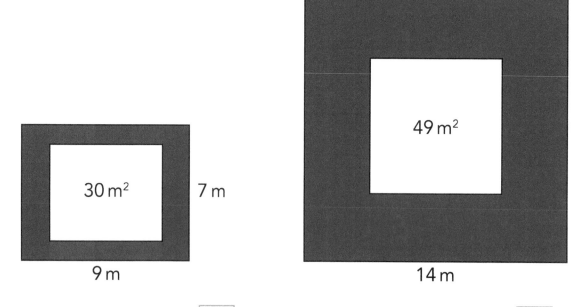

(a) Area of shaded part = ☐ m²   (b) Area of shaded part = ☐ m²

👪 Choosing the correct unit of measure

Together, talk about items in and around your house that you might calculate the area of. When discussing these, think carefully about the unit of measure you would measure in and use as part of your calculation. Ask your child to make a list of things they would measure in metres and others they would measure in centimetres.

9.1 Converting between kilometres and metres

Convert between units of measure

1. Draw lines to match the related measures.

8.4 km	802 000 m
1560 m	4.05 km
802 km	8400 m
4050 m	6030 m
6.03 km	2.01 km
2010 m	1.56 km

2. Fill in the boxes.

(a) 1.6 km = ☐ m

(b) 7 km + 37 m = ☐ km

(c) 3000 km + $\frac{1}{20}$ km = ☐ m

(d) 750 m + 450 m = ☐ km

(e) 9 km + ☐ m = 9.5 km

(f) 3.8 km − $\frac{3}{5}$ km = ☐ m

👪 Choosing the correct unit of measure

Together, talk about when you might use metres or kilometres, for example on a car journey or measuring how tall someone is. When discussing these, think carefully about the unit of measure you would measure in. Ask your child to make a list of things they would measure in metres and others they would measure in kilometres. Would they say the height of a door was 0.002 km? How would they convert this to metres?

9.2 Perimeters of rectangles and squares (1)

Calculate the perimeter of rectangles and squares

1. Fill in the table below, showing possible dimensions for these rectangles.

A: perimeter = 40 cm

B: perimeter = 36 cm

C: perimeter = 30 cm

D: (length = width)

	Perimeter (cm)	Length (cm)	Width (cm)
A	40 cm		
B	36 cm		
C	30 cm		
D	44 cm		

2. The perimeter of a shape is 60 cm. If it is a square, what must the length, width and area be? If the shape is a rectangle, what might the length, width and area be?

If it was a ...	Length	Width	Area
square			
rectangle			

Exploring patterns and relationships

Start with a 2 cm by 2 cm square and calculate the area and perimeter. Then move onto a 3 cm by 3 cm square, again calculating the area and perimeter. Do the same from squares up to 10 cm by 10 cm. Ask your child to look at their calculations. What patterns can they see?

9.3 Perimeters of rectangles and squares (2)

Calculate the perimeter of rectangles and squares

1. Solve the following problems. Use the space below to show your working.

   (a) The area of two identical rectangles, when placed together horizontally, is 216 m². The length of one rectangle is 12 m.

   [blank working box]

   (i) What is the width of the combined rectangle? [   ]

   (ii) What is the area of one rectangle? [   ]

   (iii) What is the perimeter of the new combined rectangle? [   ]

   (iv) Does the perimeter change if the rectangles

   are joined with their long sides together? _____

   (b) Ava has 20 sticks measuring 5 cm each. She arranges all the sticks to form a rectangle.

   [blank working box]

   (i) Find three different ways Ava could create the rectangle.

   (ii) What is the perimeter of these rectangles? [   ]

   (iii) Does the perimeter measurement change? Why? _____

   (iv) What is the area of each rectangle? [   ]

👪 Conjecture

Discuss this following statement together. Ask your child to use mathematical drawings to explain their decision.

The perimeter of a rectangle, when measured in whole centimetres or metres, will always have an even number of units.

9.4 Perimeters and areas of rectilinear shapes

Calculate the perimeters and areas of rectilinear shapes

1. Calculate the perimeter and area of these shapes.

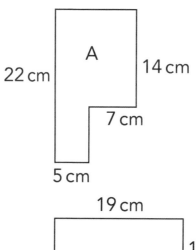

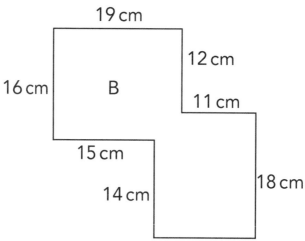

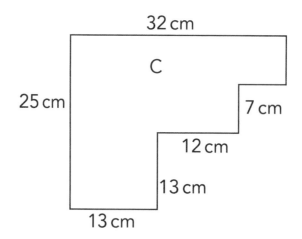

Shape A: area = _____ perimeter = _____

Shape B: area = _____ perimeter = _____

Shape C: area = _____ perimeter = _____

2. The area of a rectilinear shape is 240 cm. Its perimeter is 68 cm. Use the space below to draw a shape that matches these properties.

Working together

Ask your child to draw different shapes that have an area of 360 cm^2. How many different shapes can they generate? Which rectilinear shape would create the largest perimeter?

Describe positions and movements on a 2-D grid

1. Fill in the spaces, showing the position of each landmark.

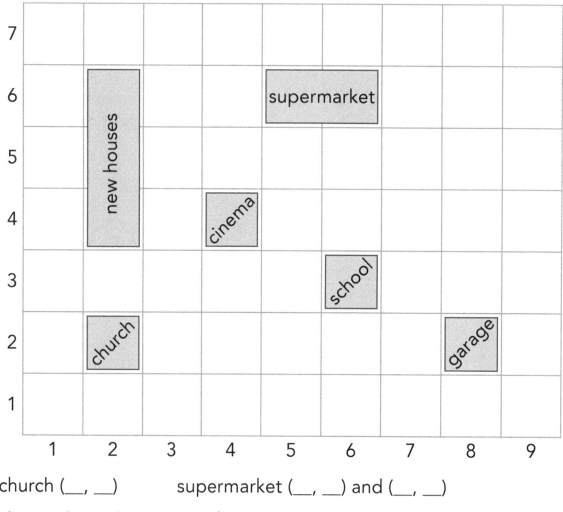

(a) church (__, __)   supermarket (__, __) and (__, __)

   cinema (__, __)   new houses (__, __) and (__, __) and (__, __)

   school (__, __)   garage (__, __)

(b) The council want to build a new road linking the new houses and the school. What is the shortest route they could take?

_____

2. Add these onto the grid above.

(a) local shop (8, 5)   restaurant (5, 4)   play park (4, 2) and (5, 3)

(b) Do you now have to change the route of your new road?

_____

👪 Garden grid

Using a similar-sized grid to the one above, ask your child to create a grid showing the approximate position of items in a garden or outside space that you know.

9.6 Solving problems involving time and money (1)

Solve problems involving time

1. Fill in the times using the clues given.

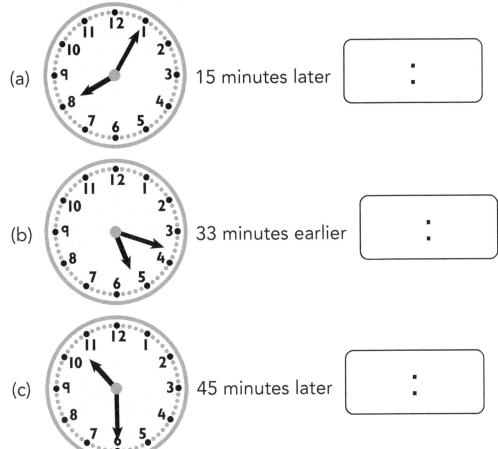

(a) 15 minutes later ⬚ : ⬚

(b) 33 minutes earlier ⬚ : ⬚

(c) 45 minutes later ⬚ : ⬚

2. Calculate.

Sam runs a full marathon (42 kilometres) in 180 minutes. If Sam's speed is always the same, how many metres does he run in each hour?

👪 Analogue and digital

Look around your home with your child to find any clocks, for example on a smartphone, the front of an oven, a wall clock. Ask your child to say whether it shows analogue or digital format. For all digital displays, help them to draw an analogue clockface and draw on the time it is showing and vice versa for analogue clocks. How many clocks used the 24-hour format?

9.7 Solving problems involving time and money (2)

Solve problems involving money and time

1. Fill in the boxes. Give your answers as whole numbers, fractions or decimals.

   (a) 705p = [ ] pounds   (b) 87p = [ ] pounds

   (c) £1.59 + 50p = [ ] pounds

   (d) Pencils cost 50p each. A teacher buys one for each
       of the 30 pupils in her class. How much does she spend?

       [ ] pence = [ ] pounds

   (e) Ben has £1.61 to spend on sweets. He decides to
       use an equal amount of this money every day
       for one week. How much can he spend every day?   [ ] pounds

   (f) Anab gets £3.25 a week pocket money.
       How much money does he get a month?   [ ] pounds

2. Calculate.
   (a) Emma gets paid £189 a week for her job as a dog walker. How much
       money does she get paid:

       (i) a day? [ ]   (ii) in half a year? [ ]

       (iii) in a whole year? [ ]

   (b) A local shop is open for 9 hours a day, 7 days a week. Customers spend
       £33 each hour. How much do customers spend in:

       (i) a day? [ ]   (ii) a week? [ ]

       The shop decides to extend their opening hours by 1 extra hour a day.
       How much extra money would customers spend:

       (iii) a day? [ ]   (iv) a week? [ ]

👪 Food shopping

Work together on this problem (round all figures to the nearest whole pound before
dividing or multiplying). How much do you spend as a household on food shopping a
week? Taking this amount, how much do you spend a day? How much do you spend
a year?

If the household reduced its weekly spending by £9 a week, what would the total
monthly and year spend be? How much would the household save?

Use multiplication and division to solve rate problems

1. Complete the table below.

   The table shows the work rate of children in Class A completing their maths tests. They all had 8 minutes to complete as many questions as possible.

	Child A	Child B	Child C	Child D	Child E	Child F
Questions answered	72	96			64	
Work rate (questions per minute)			13	19		14

2. Complete the table below.

   In a cake baking competition, contestants had to bake as many cakes as they could in a set time. The judges checked on the bakers at different points throughout the competition. The table below records their findings.

	Baker A	Baker B	Baker C	Baker D	Baker E
Time seen	7 minutes	9 minutes	4 minutes	11 minutes	14 minutes
Cakes baked	98		56		84
Work rate (cakes per minute)		12		6	

   (a) Who has the best work rate? _____

   (b) In 30 minutes, how many cakes would be baked by:

   Baker A = ☐   Baker B = ☐   Baker C = ☐

   Baker D = ☐   Baker E = ☐

   (c) How many minutes would it take for Baker C to bake 182 cakes? ☐

   (d) After 10 minutes, who would have baked the most cakes – Baker A or Baker C? ☐

👪 Jumping time

With your child, count how many jumps each of you can do in 2 minutes. What are your 'jumps per minute' rates? How many jumps would each of you do in 10 minutes?

Use multiplication and division to solve rate problems

1. Draw a tree diagram for each question.

   (a) A café is open for 9 hours a day. On Monday they make 189 cups of tea. How many did they make each hour?

        On Tuesday they make 9 more cups of tea an hour. How many do they make in total on Tuesday?

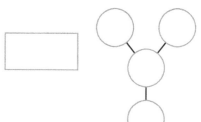

   (b) A delivery company can deliver 29 parcels an hour. Using this rate, how many parcels do they deliver in 12 hours?

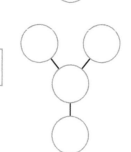

        On a busy day, the company have 408 parcels to deliver over their 12-hour working day. What will their hourly work rate need to be to meet this need?

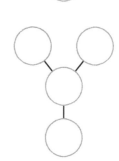

   (c) Three cyclists are taking part in a sponsored cycle for charity for 3 hours. Between them, they must cycle 54 km. Cyclist A cycles 6 km every hour. How far do Cyclists B and C need to cycle in each hour to meet their 54 km target?

Reading challenge

Together, read a short book out loud. How long does it take your child to read 10 pages? What is their 'reading rate' (pages per minute)? Calculate how many pages they could read if they carried on at the same rate for 100 minutes.

Calculate using mixed operations and brackets

1. True or false? Put a ✓ for true and a ✗ for false in each box.

   (a) $5 \times 12 \div 4 = 120 \div 3 - 25$ . . . . . . ☐

   (b) $63 - 18 \div 6 = 64 - 140 \div 20$ . . . . ☐

   (c) $110 - 54 \times 2 = 5 \times 16 - 13 \times 6$ . . ☐

   (d) $56 \div 28 \div 2 = 200 \div 10 - 2$ . . . . . ☐

2. Using only the numbers 4, 6, 8 and 2, create number sentences using the four operations and brackets to make the following totals (they might not all be possible).

   2   3   4   5   6   10

   ☐

 **Many possibilities**

Ask your child to explain the following: $6 + 12 \div 4 \times 8 = 30$ and $(44 + 6) \div 5 \times 3 = 30$. Can your child explain why brackets are not needed in the first sentence but are needed in the second?

Calculate using mixed operations and brackets

1. Use order of operations to help you solve these problems. Write the number sentence for each step.

   (a) $42 \times 12 - 945 \div 9$

   (b) $83 + 61 \times 8 - 105$

   (c) $785 \div 5 + 409 - 499$

   (d) $1000 + 684 \div 4 - 702$

2. Look at the calculations below. Have they been calculated correctly? If not, correct them.

   (a) $875 + 680 \div 5 - 99 = 212$

   (b) $648 \div 24 + 59 \times 3 = 258$

   (c) $909 + 99 \div 3 \times 8 = 1173$

   (d) Add 4 times 1554 divided by 42 to 723 and the answer is 871

👪 Many possibilities

Working together, create number sentences which involve multiple operations and brackets to achieve the answer of 105. How many different ways can your child find? Is it possible using only single-digit numbers?

Calculate using mixed operations and brackets

1. Put the calculations in 3 steps into one number sentence with mixed operations and then calculate.

   (a) $62 \times 17 = a$
   $88 + a = b$
   $b - 1058 = c$

   Number sentence: _____

   Value of a is [ ]

   Value of b is [ ]

   Value of c is [ ]

   (b) $26 \times 49 = a$
   $1086 \div 6 = b$
   $a + b = c$

   Number sentence: _____

   Value of a is [ ]

   Value of b is [ ]

   Value of c is [ ]

   (c) $38 \div 38 = a$
   $38 \times a = b$
   $b - 38 = c$

   Number sentence: _____

   Value of a is [ ]

   Value of b is [ ]

   Value of c is [ ]

   (d) $26 \times 5 = a$
   $a + 1005 = b$
   $42 \times b = c$

   Number sentence: _____

   Value of a is [ ]

   Value of b is [ ]

   Value of c is [ ]

2. Solve this.
   A wedding photographer took 483 photos in the morning; in the afternoon, she took 450 fewer photos than in the morning.

   (a) How many photos did the photographer take in total throughout the day? [ ]

   (b) The photographer can edit 39 photos an hour. After editing for 12 hours, how many more photos are not yet edited? [ ]

   (c) How many hours will this take? [ ]

   (d) Write the number sentence for this calculation. _____

Mathematical stories

Using the calculation below, ask your child to make up a real-life context for the mathematics. Check that the operations are included correctly.

$5 \times (160 \div 32) + (14 \times 9)$

10.6 Solving calculation questions in 3 steps (4)

Calculate using mixed operations and brackets

1. For each calculation, write out the number sentences in the order that you would calculate. The first has been started for you.

   (a) $278 + [(54 \times 18) \div 6]$

   First: $54 \times 18 =$ _____

   Next: _____

   Then: _____

   (b) $1000 - [318 - (19 \times 14)] =$ _____

   First: _____

   Next: _____

   Then: _____

   (c) $[43 + (29 \times 23)] \div 5 \times 7 =$ _____

   First: _____

   Next: _____

   Then: _____

   After that: _____

   (d) $551 + [(58 \times 19) \div 2] =$ _____

   First: _____

   Next: _____

   Then: _____

2. Calculate the following problems using order of operations.

   (a) $[600 + (264 \div 11)] \div 48$

   (b) $1056 \div [288 \div (216 \div 12)]$

   (c) $24 \times [1000 \div (100 \div 10)]$

   (d) $49 + [62 \times (32 \times 32)] - 497$

👪 Summary

Discuss the mathematical rules associated to calculating using the four operations and brackets. Together, create a poster to explain these rules to someone else – maybe a family member – including some of your child's own examples of calculations for them to complete.

10.7 Working forwards

Calculate using mixed operations

1. Fill in the boxes and calculate. Write the mixed operation number sentence underneath.

(a) 650 $\xrightarrow{\div 10}$ ☐ $\xrightarrow{\times 8}$ ☐ $\xrightarrow{\div 2}$ ☐

Number sentence: _____

(b) 500 $\xrightarrow{\times 3}$ ☐ $\xrightarrow{-750}$ ☐ $\xrightarrow{\div 10}$ ☐

Number sentence: _____

(c) 2573 $\xrightarrow{+499}$ ☐ $\xrightarrow{-500}$ ☐ $\xrightarrow{\div 4}$ ☐

Number sentence: _____

(d) Multiply 58 by 13, then subtract 137 before dividing by 2.

Number sentence: _____

2. Complete these problems.

(a) First find the product of 53 and 34, before subtracting the quotient of 1335 divided by 15.

Number sentence: _____

(b) An aeroplane can hold 450 passengers on its flight.
164 passengers get on first, followed by 3 lots of 90 passengers.
How many more can fit onto the plane?

Number sentence: _____

(c) A school kitchen has 592 meals to serve in an hour. In the first half an hour they are able to serve 15 classes of 30 children. In the following 15 minutes they serve the total number of meals divided by 37. How many children are left to serve?

Number sentence: _____

4 card operations

On 10 small pieces of paper, write an operation on each (for example, × 6 or ÷ 5). Agree on a 3-digit start number and select four of the cards to show the process of calculation. Can your child work forwards and complete all four calculations? Repeat with different cards.

10.8 Working backwards

Calculate using mixed operations

1. Fill in the boxes and calculate. Write the mixed operation number sentence underneath.

(a) ⬚ —— ÷ 10 ——→ ⬚ —— × 8 ——→ ⬚ —— ÷ 2 ——→ 300

   Number sentence: _____

(b) ⬚ —— + 579 ——→ ⬚ —— × 3 ——→ ⬚ —— ÷ 15 ——→ 155

   Number sentence: _____

(c) ⬚ —— − 999 ——→ ⬚ —— ÷ 14 ——→ ⬚ —— − 199 ——→ 11

   Number sentence: _____

(d) ⬚ —— ÷ 10 ——→ ⬚ —— × 8 ——→ ⬚ —— ÷ 2 ——→ 120

   Number sentence: _____

2. Complete these problems.
   (a) After a number has been divided by 7, 38 is added and the total is multiplied by 4, the number is 224.

   Number sentence: _____

   (b) A number is multiplied by 7 after being divided by 14. The number is 314. What is the number?

   Number sentence: _____

   (c) A theatre makes £9670 after a performance. This profit was made up from: £1912 made on drinks; £3862 made on food; and 48 brochures sold at £4.00 each. The rest was on entry tickets. How much was taken for entry tickets?

   Number sentence: _____

👪 4 card operations

Repeat the previous '4 card operations' activity. On 10 small pieces of paper, write an operation on each (for example, × 6 or ÷ 5). Agree on a number that will be the result of your child's calculations and select four of the cards to show the process of calculation. Can your child work backwards and complete all four calculations? Repeat with different cards.

Solve word problems involving mixed operations

1. Read these problems carefully and calculate.

   (a) What is the difference between the product of 299 and 21 and the product of 598 and 11?

   (b) What is the sum of the product of 83 and 62, and minus 1001 subtract 899?

   (c) What is the quotient if the dividend is the product of 26 and 63, and the divisor is a tenth of 130?

   (d) What is the product of a third of 66 and an eighth of 112?

2. Match up two statements that, when calculated, give the answer of 124. One has been completed for you.

   143 + 255   divided by 15

   1284 + 576   divide by 15 and add 101

   28 × 11   subtract two hundred and seventy-four

   69 × 5   minus 184

🚶 Mathematical stories

Using the calculation below, ask your child to make up a real-life context for the number sentence. Make sure that they take note of the brackets and operations.

1964 − (132 × 10) + 42

10.10 Word calculation problems (2)

Solve word problems involving mixed operations

1. Read these problems carefully and calculate. Write the number sentence for each problem.

   (a) The difference between numbers A and B is the product of 31 and 123.

      (i) If number A is 498, what is number B?

         Number sentence: _____

      (ii) If number B is 4339, what is number A?

         Number sentence: _____

   (b) The product of 61 and 24 is the same as 12 multiplied by a different number. What is this number?

      Number sentence: _____

   (c) The sum of number A and number B is 1935.

      (i) If number A is the product of 74 and 15, what is number B?

         Number sentence: _____

      (ii) If number B is the quotient of 2000 by 25, what is number A?

         Number sentence: _____

2. The sum of A and B is multiplied by C to give D. Complete this chart to show the calculations and write the number sentence for each problem.

	Number A	Number B	Number C	Number D
1	102	58	12	
2	178		18	4644
3		199	13	5278
4	1243	4295	0	

Problem 1: _____

Problem 2: _____

Problem 3: _____

Problem 4: _____

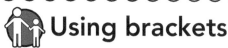 Using brackets

Look at Question 2 again. Discuss with your child where brackets have been used and why they were needed.

10.11 Laws of operations (1)

Use the commutative and associate laws to calculate efficiently

1. Fill in the boxes to show different ways to do each calculation using the law of operations.

   (a) 1256 + 295 + 173 = ☐

      ☐ + ☐ + ☐ = ☐

      or ☐ + ☐ + ☐ = ☐

   (b) 52 × 16 × 12 = ☐

      ☐ × ☐ × ☐ = ☐

      or ☐ × ☐ × ☐ = ☐

   (c) 3872 + 1243 − 751 = ☐ − ☐

   (d) A × B × C = ☐ × ☐

   (e) ☐ × 67 = ☐ × 24

   (f) 99 + 99 + 99 + 99 = ☐ + ☐

2. Simplify and calculate.

   (a) 253 + 512 + 432

   (b) 89 × (12 × 24)

   (c) 9 × 12 × 14

   (d) 123 + 456 + 789

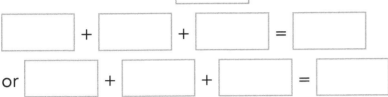 Simplifying

Together, decide how to simplify the calculation below. There might be more than one way.

3 × 10 × 25 × 4

Use the commutative and associate laws to calculate efficiently

1. Simplify and calculate.

    (a) $6 \times 72 \times 2 \times 10$

    (b) $(147 + 123) \times 12$

    (c) $24 \times 14 \times 10$

    (d) $224 \times 14 \times 5 \times 2$

    (e) $2500 + 190 + 750$

    (f) $125 \times 3 \times 72$

2. True or false? Put a ✓ for true and a ✗ for false in each box.

    (a) $23 + 83 \times 4 = 23 + (83 \times 4)$ . . . . . ☐

    (b) $264 - 83 - 54 = 264 - (83 + 54)$. ☐

    (c) $67 \times 12 \times 23 = 67 \times 35$ . . . . ☐

    (d) $(85 + 62) \times 6 = 85 + (62 \times 6)$ . . . . ☐

    (e) $23 \times 15 = 23 \times 9 + 23 \times 6$ . . . . . . ☐

    (f) $12 \times 100 = 120 \times 10$ . . . . ☐

👪 Why is it not the same?

Working together, look at the problem below and discuss why the two statements are *not* equal. Can your child explain this mathematically using their knowledge of the law of operations?

$(85 + 62) \times 6 = 85 + (62 \times 6)$

Use the distributive law to calculate efficiently

1. In your own words, explain what mathematicians mean by 'distributive law'.

_____

_____

_____

_____

_____

2. Simplify each calculation below and find the answer.

(a) $45 \times 41 + 45 \times 9$

(b) $74 \times 12 - 12 \times 32$

(c) $(43 + 25) \times 16$

(d) $19 \times 15 - 15 \times 6 + 15 \times 7$

(e) $43 \times (16 \times 21)$

(f) $(61 - 43) \times 8$

Important concept

The 'distributive law' of multiplication over addition means that where an addition in brackets is multiplied by a number, each part of that addition can be written as a separate multiplication and then those two products added together. For example, $12 \times (8 + 9)$ can be calculated as $(12 \times 8) + (12 \times 9)$. Together, make up another example.

Use laws of operations to calculate efficiently

1. Use your preferred method to simplify and calculate the following.

(a) 199 × 42

(b) 115 × 39

(c) 201 × 63

(d) 113 × 87

(e) 49 × 209

(f) 50 × 35

2. Solve these application problems.

(a) At the beginning of a day, with an empty till, a florist has 48 bunches of tulips and 58 bunches of roses. Each bunch costs £4. If the florist sells all the bunches, how much money will they have?

Number sentence: _____

(b) Toy store A sells 102 dolls priced at £8.00 each. Toy store B sells 51 of the same doll at a reduced price of £4.00. Which toy store makes the most money?

Number sentence: _____

(c) A baker needs 150 g of butter and 205 g of flour to make one cake. The baker needs to make 37 cakes; how much of each ingredient will they need?

Number sentence: _____

👪 Why does it work?

Together, explore the problem below. It has been simplified, but can your child explain mathematically why it is still correct? What are the relationships that they recognise in the calculation? Can they think of their own similar example?

10 × 36 + 10 × 72 can be simplified to 30 × 36, but why?

10.15 Problem solving using four operations (1)

Solve problems using four operations

Solve these problems. Show your calculation for each problem.

(a) A zookeeper has 3978 items of fruit to feed the monkeys. This fruit should have lasted him 26 days, but the monkeys were overfed and it only lasted 18 days. How much more food did the monkeys eat each day?

Calculation:_____

(b) A long-distance cyclist is taking part in a 1040 km cycle across the country. He has 20 days to complete the race. He actually completes it in half the given time.

(i) How far was he supposed to travel each day?

Calculation: _____

(ii) How far did he actually travel each day?

Calculation: _____

(c) A shop has 1 hour to put 1560 items of stock on the shelves before it opens.

(i) How many items should be put on shelves in each minute?

Calculation: _____

(ii) They actually manage to complete the restocking in half the time. How many items do they actually restock in each minute?

Calculation: _____

(d) A bakery needs to make 1200 cakes for a special party. Baker A could complete this order in 30 hours, but Baker B could complete this order in 25 hours. How many more cakes can Baker B make an hour compared to Baker A?

Calculation: _____

~~~~~~

👪 Create your own

The problems above all include the previous learning on 'Work rate'. Ask your child to generate their own questions that include calculating work rate similar to the problems explored on this page. These problems will be given to a friend at school to solve.

10.16 Problem solving using four operations (2)

Solve problems using four operations

1. Write two different methods with mixed operations and then calculate.

(a) A racing car travels at a speed of 120 km per hour along a straight road. From start to finish it, completes the race in 4 minutes. How long was the race track?

Method 1	Method 2

(b) Bryony is reading a book. It has 690 pages in it. She hopes to complete the book in 30 days. To achieve this, how many pages will she need to read a day?

Method 1	Method 2

(c) The world record for pairing socks is 31 pairs in a minute. Michael is planning to beat this record. After 4 minutes he has managed to pair 116 socks. Has he beaten the world record?

Method 1	Method 2

(d) A gardener can cut 80 square metres of grass every 4 minutes. They have 720 square metres to cut so decide to employ another gardener to help; she works at the same rate. Together, how long will it take them to cut all the grass?

Method 1	Method 2

Create your own

Help your child to create a word problem for each of the four operations. They can use the previous learning about work rate and multi-step problems to guide them when creating. Ask them to write out the methods that could be used to solve each of the problems afterwards.

10.17 Problem solving using four operations (3)

Solve problems using four operations

1. Read these problems carefully and calculate.

   (a) Edward wants to buy a new football for £6.00, a board game for £3.50 and a computer game for £36.75.

   (i)   How much money does Edward need in total?

   (ii) He gets £4.50 a week pocket money. For how many weeks will he need to save?

   (iii) When he has enough money to buy the items, will he have any pocket money left over?

   (b) A café has 32 tables. A quarter of the tables are for 2 people and the rest are for 4 people.

   (i)   How many chairs does the café need for their 32 tables?

   (ii) How many tables sit 2 people?

   (iii) How many tables sit 4 people?

   (c) Henry is given £48 for his birthday. He buys 6 new books and 3 new T-shirts. The books cost £4.00 each and the T-shirts cost £3.50.

   (i)   How much does he spend on these items in total?

   (ii) Will Henry have any money left over?

   (iii) Could Henry buy more books and T-shirts? If so, how many of each?

   (d) Aisha has 16 red beads, a quarter as many green beads as red beads and half as many blue beads as green beads. How many beads does she have in total?

Create your own

Together, make up a problem that uses the number sentence below. Make sure your child checks the operations and watches out for the brackets. Then ask your child to make up another problem with the same numbers.

(£18 + £21 + £11) ÷ 4

Solve problems using four operations

1. Calculate the following multi-step problems.
   (a) A family are travelling to see their relatives a long way from their home. They have to travel 438 km. In the first hour they travel 74 km. In the second hour, they travelled 64 km. For every hour after, they travelled 60 km an hour.

   (i) How many kilometres did they travel in the first 2 hours?

   (ii) How many hours did it take them to complete the journey?

   (b) DVDs are sold for £6.50 each. However, there is a special offer and you can buy 3 for £15.

   (i) How much would 3 DVDs at full price be?

   (ii) How much money do you save with the special offer?

   (iii) Write the number sentence showing how you calculated Question (ii).

   _____

   (c) The area of a rectangular sports field is 216 m². The width of the field is 12 m. The groundkeeper needs to know the perimeter – what is it?

   Calculation: _____

   (d) Daniel bought 4 cookies and 3 packets of crisps. He spent £4.80. The cookies are $\frac{3}{4}$ the price of the crisps.

   (i) How much is a packet of crisps?

   (ii) How much is a cookie?

   (iii) What is the total cost of 3 packets of crisps?

   (iv) What is the total cost of the 4 cookies?

Learning summary

Together, create a poster which explains and shows how your child solved the problems above. Be clear with the method so that someone else can read it and understand how it was calculated.

Answers

1.1 Warm up revision

Q1 (a) 600   (b) 590   (c) 501
     (d) 601   (e) 601   (f) 557

Q2 (a) 1010
      1010
      566 and 1010
      1110

     (b) 816   (c) 409
        915   509
        918   249
        164   228

Patterns:

SET A: Tens and ones numbers equal 10

SET B: Tens numbers all equal 11, resulting in the tens number being '1' in the sum

SET C: The ones in the subtrahend is one more that in the minuend, resulting in bridging.

1.2 Multiplication tables up to 12 × 12

Q1

x	7	12	8	5	6	9	11	4	10
6	42	72	48	30	36	54	66	24	60
7	49	84	56	35	42	63	77	28	70
8	56	96	64	40	48	72	88	32	80
9	63	108	72	45	54	81	99	36	90
12	84	144	96	60	72	108	132	48	120

   (a) They represent doubles and halves respectively.
   (b) Pairs: 3 and 6, 6 and 12, or others.

Q2 (a) 54   (b) 63   (c) 7
     (d) 8   (e) 72

Q3 (a) 105   (b) 117
     (c) 8 × 10 + 8 × 4

1.3 Multiplication and division (1)

Q1

Racer	Distance (m)	Time (s)	Speed (m/s)
Yellow	144	9	**16**
Red	126	9	**14**
Blue	**128**	8	16
Green	126	**14**	9

Q2 Pupils should show the correct calculation using the column method.
     (a) 456   (b) 581   (c) 536
     (d) 65   (e) 88   (f) 9

1.4 Multiplication and division (2)

Q1 (Pupils should present their answers using the column method)
     (a) 2086   (b) 123
     (c) 2436   (d) 608

Q2

Item	Number of fruits in each box	Number of boxes	Total items
Apples	9	110	990
Strawberries	7	280	**1960**
Bananas	6	**151**	906
Pears	8	399	**3192**
Melons	5	**242**	1210

1.5 Problem solving (1)

Q1 88 minutes

Q2 (a) Mike writes 315 words. Pupils should provide 2 methods to show how they reach this answer.
     (b) Finn runs 1064 metres. Kayo runs 760 metres before her injury. Pupils should provide 2 methods to show how they reach the answers.

1.6 Problem solving (2)

Q1 (a) 460 ÷ 4 = 115,
        115 + 460 = 575
     (b) 460 × 3 = 1380,
        1380 + 460 = 1840,
        1840 ÷ 2 = 920
     (c) 460 + 115 = 575,
        1380 − 575 = 805,
        805 ÷ 5 = 161

Q2 (a) 69
     (b) Farmer B = 64
        Farmer C = 96
        Total number of animals = 192

1.7 Fractions

Q1 $\frac{1}{2}$ or $\frac{2}{4}$

     $\frac{1}{4}$ or $\frac{2}{8}$

     $\frac{1}{4}$ or $\frac{2}{8}$

Q2 Emma gets 5 cookies.
     Paul gets 5 cookies
     Mike gets 5 cookies
     George gets 5 cookies.

2.1 Knowing numbers beyond 1000 (1)

Q1

Number	Number in words
5005	Five thousand and five
11640	Eleven thousand six hundred and forty
28702	**Twenty eight thousand seven hundred and two**
10004	**Ten thousand and four**
12037	Twelve thousand and thirty-seven
14321	**Fourteen thousand three hundred and twenty one**

Q2 (a) Smallest possible number: 1599; Largest possible number: 9960
     (b) 1236, 2448, 3648, 3624 – or other variations like these.
     (c) Any four from 5790, 5970, 7590, 9750, 7950, 9570

2.2 Knowing numbers beyond 1000 (2)

Q1 9836; 8693; 8396; 6893; 6398
     (a) (i)   False
         (ii)   True
         (iii)   True
         (iv)   False
         (v)   True 3689

Q2 (a) 3045
     (b) 9754
     (c) Many possibilities
     (d) A number that has 7 or 9 thousands that includes the digits given.
     (e) For example, 9750 or 9543

2.3 Rounding numbers to the nearest 10, 100 and 1000

Q1 (a) Multiples of thousand

6000	6777	**7000** ✓
8000 ✓	8163	**9000**
1000 ✓	1006	**2000**
0	964	**1000** ✓

(b) Multiple of hundred

6700	6777	**6800** ✓
8100	8163	**8200** ✓
1000 ✓	1006	**1100**
900	964	**1000** ✓

(c) Multiple of ten

6770	6777	**6780** ✓
8160 ✓	8163	**8170**
1000	1006	**1010** ✓
960 ✓	964	**970**

Q2

Number	Nearest thousand	Nearest hundred	Nearest ten
7652	8000	7 700	7 650
5946	6000	5 900	5950
9040	9000	9 000	9 040
4265, 4266, 4267, 4268, 4269, 4270, 4271, 4272, 4273, 4 274	4000	4 300	4 270

Q3 Rounded to the nearest thousand: Town A (5000) Rounded to the nearest hundred: Town A (4500) and Town B (4500) are the same.

2.4 Addition with 4-digit numbers (1)

Q1 (a) 5890 **(b)** 6123
(c) 10 000 **(d)** 9964

Q2 (a) 6912 **(b)** 9218 **(c)** 1100

2.5 Addition with 4-digit numbers (2)

Q1 All problems should be calculated using the column method.
(a) 10 991 **(b)** 10 000
(c) 4713 **(d)** 12 243

Q2 (a)
```
  8 4 2 1
+ 1 3 9 9
---------
  9 8 2 0
```

(b)
```
  1 2 3 4
+ 5 6 7 8
---------
  6 9 1 2
```

(c)
```
    5 6 7 8
+   5 6 7 8
-----------
  1 1 3 5 6
```

(d)
```
    9 2 1 4
+     8 8 5
-----------
  1 0 0 9 9
```

2.6 Subtraction with 4-digit numbers (1)

Q1 (a) 3111 **(b)** 5283
(c) 4655 **(d)** 4094

Q2 (a) 4588 **(b)** 3545 **(c)** 9506

2.7 Subtraction with 4-digit numbers (2)

Q1 (a) 1951 **(b)** 3109
(c) 103 **(d)** 5027

Q2 (a)
```
  ⁵6̶ ¹3̶ 8 4
-   5 4 7 3
-----------
      9 1 1
```

(b)
```
  9 9 9 9
- 8 7 6 5
---------
  1 2 3 4
```

(c)
```
  ⁶7̶ ¹⁰1̶ ¹³4̶ ¹3
-   3  2  5  4
--------------
    3  8  8  9
```

(d)
```
  ⁸9̶ ⁹0̶ ⁹0̶ ¹0
-   5  0  8  3
-------------
    3  9  1  7
```

2.8 Estimating and checking answers using inverse operations

Q1 (a) 1106 − 225 = 881
225 + 881 = **1106**
1106 − **881** = **225**
881 + 225 = **1106**

(b) 5237 − 1555 = 3682
3682 + **1555** = **5237**
1555 + 3682 = 5237
1555 = **5237** − **3682**

(c) 5726 − 1107 = **4619**
1107 = 5726 − **4619**
5726 = 1107 + **4619**
5726 = **4619** + **1107**

Q2 (a) 3546 + 1088 = 4634
(b) 4446 − 1074 = 3372
(c) 4250 = 290 + 3960
(d) 4612 − 1062 = 3550
(e) 6499 = 1850 + 4649
(f) 4256 + 991 = 5247
(g) 9632 + 1232 − 7267 = 3597
(h) 8812 − 4565 − 566 = 3681

Q3 (a) 6100, 6066
(b) 4100, 4134

3.1 Multiplying whole tens by a 2-digit number

Q1 (a) 740 **(b)** 720
(c) 11 200 **(d)** 4500
(e) 33 000 **(f)** 33 000

Q2 (a) $15 \times 2 = 30$
$150 \times 2 = 300$
$15 \times 20 = 300$
$150 \times 200 = 30\,000$
(b) $60 \times 5 = 300$
$6 \times 500 = 3000$
$60 \times 50 = 3000$
$600 \times 5 = 3000$
(c) $72 \times 3 = 216$
$720 \times 3 = 2160$
$72 \times 30 = 2160$
$72 \times 300 = 21\,600$

3.2 Multiplying a 2-digit number by a 2-digit number (1)

Q1 (a) 704 **(b)** 7533
(c) 2430 **(d)** 1971
(e) 2156 **(f)** 3015

Q2 (a) 1674
(b) $35 \times 35 = 1225$
(c) 168
(d) 39

3.3 Multiplying a 2-digit number by a 2-digit number (2)

Q1 (a) 585   (b) 1071
(c) 962   (d) 1504

Q2 (a)
$$
\begin{array}{r}
5\ 7 \\
\times\ \ 3\ 6 \\
\hline
3\ 4\ 2 \\
1\ 7\ 1\ 0 \\
\hline
2\ 0\ 5\ 2
\end{array}
$$
(b)
$$
\begin{array}{r}
1\ 4 \\
\times\ \ 1\ 8 \\
\hline
1\ 1\ 2 \\
1\ 4\ 0 \\
\hline
2\ 5\ 2
\end{array}
$$

(b) Pupil explanations should include the incorrect use of place holder (0) and incorrect calculation.

3.4 Multiplying a 3-digit number by a 2-digit number (1)

Q1 Answer should be presented using the column method.
(a) 3552   (b) 4242
(c) 8652   (d) 33 615
(e) 13 104   (f) 13 622

Q2
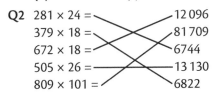

$281 \times 24 =$ 12 096
$379 \times 18 =$ 81 709
$672 \times 18 =$ 6744
$505 \times 26 =$ 13 130
$809 \times 101 =$ 6822

3.5 Multiplying a 3-digit number by a 2-digit number (2)

Q1 (a) $32 \times 4 = 128$
$32 \times 40 = 1280$
$32 \times 140 = 4480$
$320 \times 14 = 4480$
(b) $46 \times 5 = 230$
$460 \times 5 = 2300$
$46 \times 50 = 2300$
$46 \times 15 = 690$
(c) $189 \times 75 = 14\ 175$
$189 \times 750 = 141\ 750$
$1890 \times 75 = 141\ 750$

Q2 (a) 9672   (b) 22 339
(c) 16   (d) 6241

3.6 Dividing 2-digit or 3-digit numbers by tens

Q1 (a) 4r5   (b) 5r72
(c) 99   (d) 15r9
(e) 11r54   (f) 30r34

Q2

Type	Number ordered	Number in a bunch	Number of bunches and remainder
Lily	6528	30	**217 r18**
Tulip	8372	50	**167 r22**
Sunflower	**1058**	20	52 r18
Rose	7104	40	**177 r24**

3.7 Practice and exercise

Q1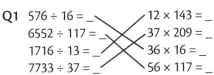
$576 \div 16 =$ _ $12 \times 143 =$ _
$6552 \div 117 =$ _ $37 \times 209 =$ _
$1716 \div 13 =$ _ $36 \times 16 =$ _
$7733 \div 37 =$ _ $56 \times 117 =$ _

Q2 (a) $972 \div 54 = 18$
(b) $66 \times 17 = 1122$
(c) $29 \times 14 = 406$
(d) $949 \div 13 = 73$

4.1 Fractions in hundredths

Q1
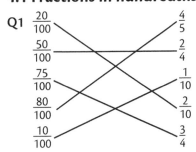

$\frac{20}{100}$ $\frac{4}{5}$
$\frac{50}{100}$ $\frac{2}{4}$
$\frac{75}{100}$ $\frac{1}{10}$
$\frac{80}{100}$ $\frac{2}{10}$
$\frac{10}{100}$ $\frac{3}{4}$

Q2

$\frac{1}{10}$ $\frac{2}{4}$ $\frac{4}{5}$

0 $\frac{7}{100}$ $\frac{2}{3}$ $\frac{70}{100}$ $\frac{89}{100}$

4.2 Addition and subtraction of fractions (1)

Q1 (a) $\frac{4}{5}$ (b) $\frac{49}{50}$
(c) 1 or $\frac{30}{30}$ (d) $\frac{79}{100}$
(e) $\frac{450}{600}$ (f) $\frac{39}{40}$

Q2 (a) $\frac{28}{34}$ (b) $\frac{13}{15}$
(c) $\frac{23}{42}$
(d) 2 and a half times
(e) $\frac{13}{25}$ (f) $\frac{15}{35}$

4.3 Addition and subtraction of fractions (2)

Q1 (a) $\frac{4}{7} + \frac{2}{7} = \frac{6}{7}$

(b) $\frac{6}{14} + \frac{5}{14} = \frac{11}{14}$

(c) $\frac{15}{20} - \frac{9}{20} = \frac{6}{20}$

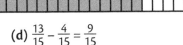

(d) $\frac{13}{15} - \frac{4}{15} = \frac{9}{15}$

Q2 (a) $\frac{6}{25}$ (b) $\frac{20}{100}$
(c) $\frac{3}{23}$ (d) $\frac{10}{30}$
(e) $\frac{1}{8}$ (f) $\frac{23}{25}$

4.4 Fun with exploration – 'fraction wall'

Q1 (a) $\frac{1}{3}, \frac{2}{x}, \frac{3}{9}$
(b) $\frac{2}{3}, \frac{4}{6}, \frac{9}{12}$
(c) $\frac{3}{4}, \frac{6}{8}, \frac{9}{12}$
(d) no equivalent fraction on the wall
(e) many possibilities
(f) many possibilities

Q2 (a) >   (b) >
(c) =   (d) <, <
(e) >, <   (f) =, =

5.1 Multiplication and multiplication tables

Q1 (a) $9 \times 7 = 63$
(b) $7 \times 9 = 63$
(c) $5 \times 6 = 30$
(d) $9 + 9 + 9 + 9 + 9 + 9 + 9 + 9 + 9 + 9 + 9 + 9 + 9 = 117$
(e) $6 + 6 + 6 + 6 + 6 + 6 + 6 + 6 + 6 + 6 + 6 + 6 + 6 + 6 + 6 = 90$
(f) $8 \times 8 = 64$

Q2 (a) $40, 8 \times 5 = 40,$
$5 \times 8 = 40, 40 \div 8 = 5,$
$40 \div 5 = 8$
(b) $42, 7 \times 6 = 42, 6 \times 7 = 42,$
$42 \div 7 = 6, 42 \div 6 = 7$

Q3 The following numbers should be circled:
12, 24, 36

Q4 $2 \times 6 = 12, 6 \times 2 = 12, 12 \div 2 = 6,$
$12 \div 6 = 2$

$2 \times 12 = 24, 12 \times 2 = 24,$
$24 \div 2 = 12, 24 \div 12 = 2$

$2 \times 18 = 36, 18 \times 2 = 36,$
$36 \div 2 = 18, 36 \div 18 = 2$

$4 \times 3 = 12, 3 \times 4 = 12, 12 \div 3 = 4,$
$12 \div 4 = 3$

$4 \times 6 = 24, 6 \times 4 = 24, 24 \div 4 = 6,$
$24 \div 6 = 4$

$4 \times 9 = 36, 9 \times 4 = 36, 36 \div 9 = 4,$
$36 \div 4 = 9$

$6 \times 6 = 36, 36 \div 6 = 6$

5.2 Relationship between addition and subtraction

Q1 (a) Many possibilities. Difference must be 198.
(b) 667
(c) Many possibilities.
(d) 697
(e) 259
(f) Sum – addend = addend;
minuend – subtrahend
= difference;
minuend – difference
= subtrahend

Q2 (a) False
(b) False. They both could be odd.
(c) False, it will be odd.
(d) False.
(e) True.
(f) True

5.3 The relationship between multiplication and division

Q1 (a) 16   (b) 38
(c) 13   (d) 92
(e) 5152   (f) 29

Q2

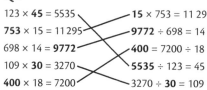

$123 \times 45 = 5535$
$753 \times 15 = 11\,295$
$698 \times 14 = 9772$
$109 \times 30 = 3270$
$400 \times 18 = 7200$
$15 \times 753 = 11\,295$
$9772 \div 698 = 14$
$400 = 7200 \div 18$
$5535 \div 123 = 45$
$3270 \div 30 = 109$

5.4 Multiplication by 2-digit numbers

Q1 (a) 98   (b) 45
(c) 484   (d) 17

Q2 (a) The product of 267×23 is a four-digit number.
(b) The digits in the product of 462×14 are all even.
(c) The product of 234×19 rounded to the nearest hundred is 4400
(d) If one factor is an odd number the product will sometimes be odd.
(e) 243 must be multiplied by 41 to make the answer as close as possible to 10 000.

5.5 Practice with fractions

Q1 $\frac{2}{8}$ or $\frac{1}{4}$ $\frac{1}{5}$
$\frac{3}{10}$ $\frac{3}{16}$

Q2 $\frac{20}{100}, \frac{1}{4}, \frac{3}{5}, \frac{6}{8}, \frac{9}{10}$

5.6 Roman numerals to 100

Q1

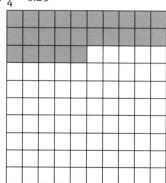

Q2 (a) $55 = LV$ (b) $90 = XL$
(c) $8 = VIII$ (d) $18 = XVIII$
(e) $28 = XXVIII$ (f) $33 = XXXIII$
(g) $101 = CI$ (h) $14 = XIV$
(i) $114 = CXIV$ (j) $200 = CC$

6.1 Decimals in life

Q1 (a) fifteen pounds and thirty pence
(b) three point four kilograms
(c) £56.83
(d) 33.6 L
(e) one metre and seven-nine centimetres
(f) 123.9p

6.2 Understanding decimals (1)

Q1 (b) $\frac{1}{4} = 0.25$

(c) $\frac{1}{2} = 0.50 = 0.5$

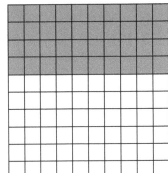

(d) $\frac{2}{4} = 0.75$

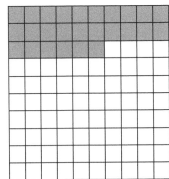

(e) $0.3 = 0.30 = \frac{30}{100}$

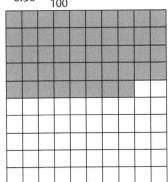

(f) $\frac{4}{10} = 0.40 = 0.4$

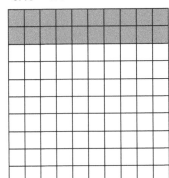

Q2 (a) $0.7 = \frac{7}{10}$ (b) $0.07 = \frac{7}{100}$

(c) $0.16 = \frac{16}{100}$ (d) $0.32 = \frac{32}{100}$

6.3 Understanding decimals (2)

Q1 (a) Count in 0.1s
0.5, 0.6, **0.7, 0.8, 0.9, 1.0**, 1.1,
1.2, 1.3, 1.4

(b) Count in 0.01s
0.93, 0.94, **0.95, 0.96, 0.97,
0.98, 0.99, 1, 1.01**

(c) Count in 0.2s
2.2, **2.4, 2.6, 2.8, 3**, 3.2, **3.4,
3.6, 3.8**

(d) Count in 0.05s
0.15, 0.2, 0.25, 0.3, 0.35, **0.4,
0.45, 0.5, 0.55**

Q2 (a) The fraction $\frac{25}{100}$ can be
written as a decimal as **0.25**
or **0.250**.

(b) A number with 42 tenths is
more than 4 (4.2).

(c) The number **0.811** consists of
8 tenths and 11 thousandths.

(d) The decimal **0.2** when
written as a fraction is $\frac{20}{100}$.

(e) There are four **tens**, six
ones, five **tenths**, three
hundredths and eight
thousandths in the number
46.538.

6.4 Understanding decimals (3)

Q1 (a) $0.876 = \mathbf{8} \times 0.1 + \mathbf{7} \times 0.01 +$
$\mathbf{6} \times 0.001$

(b) $0.054 = \mathbf{0} \times 1 + \mathbf{0} \times 0.1 +$
$\mathbf{5} \times 0.01 + \mathbf{1} \times 0.001$

(c) $2.803 = \mathbf{2} \times 1 + \mathbf{8} \times 0.1 +$
$\mathbf{0} \times 0.01 + \mathbf{3} \times 0.001$

(d) $42.51 = \mathbf{4} \times 10 + \mathbf{2} \times 1 +$
$\mathbf{5} \times 0.1 + \mathbf{1} \times 0.01$

(e) $65.703 = \mathbf{6} \times 10 + \mathbf{5} \times 1 +$
$\mathbf{7} \times 0.1 + \mathbf{3} \times 0.001$

Q2 (a) <   (b) >
(c) =   (d) >
(e) >

6.5 Understanding decimals (4)

Q1

	One decimal place	Two decimal places	Three decimal places
Mixed decimal	328.7, 1073.7	32.87, 6.50	5.405
Pure decimal	0.5	0.78	0.123, 0.004, 0.710

Q2 (a) $1\frac{5}{20} = 1.25$ (b) $\frac{25}{20} = 1.25$

(c) $\frac{32}{10} = 3.2$ (d) $4\frac{13}{25} = 4.52$

(e) $\frac{190}{100} = 1.9$ (f) $\frac{1900}{1000} = 1.9$

6.6 Understanding decimals (5)

Q1 The objects drawn to the
following lengths according to
the ruler:
(a) glue stick: 7.3 cm;
(b) rubber: 4.5 cm;
(c) pencil: 9 cm;
(d) coin: 2 cm diameter]

Q2

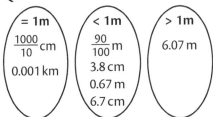

6.7 Understanding decimals (6)

Q1 (a) D   (b) A
(c) D   (d) B

Q2 25.784

6.8 Comparing decimals (1)

Q1 (a) >   (b) =
(c) >   (d) <
(e) >   (f) <

Q2 (a) 8.05, 8.15, 8.5, 8.51
(b) 0.048, 0.084, 0.408, 0.804
(c) 6.48, 6.5, 6.54, 6.84
(d) 0.7, 0.707, 0.72, 0.727
(e) 8.08, 8.8, 8.808, 8.88
(f) 3.12, 3.612, 3.62, 36.12

6.9 Comparing decimals (2)

Q1 (a) Rounding 13.4 to the nearest whole number, the result is 13.
(b) Rounding 0.12 to the nearest whole number, the result is 0.
(c) Rounding 1.04 to the nearest whole number, the result is 1.
(d) Rounding 99.06 to the nearest whole number, the result is 99.
(e) Rounding 50.7 to the nearest whole number, the result is 51.
(f) Rounding 18.0 to the nearest whole number, the result is 18.

Q2

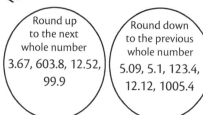

Round up to the next whole number
3.67, 603.8, 12.52, 99.9

Round down to the previous whole number
5.09, 5.1, 123.4, 12.12, 1005.4

6.10 Properties of decimals

Q1 (a) 500.01   (b) 41.1
(c) 450.1   (d) 1000.12
(e) 80.9   (f) 15.707
(g) 1000   (h) 17.1

Q2 (a) =   (b) <
(c) =   (d) <
(e) >   (f) >
(g) >   (h) >

7.1 Knowing line graphs (1)

Q1 (a) 5 and 7   (b) 7 and 9
(c) 1 and 12   (d) 9 months
(e) 1 kg

7.2 Knowing line graphs (2)

Q1 (a) Week 5
(b) Week 6
(c) Weeks 1, 2, 3, 5, 7 and 8
(d) Weeks 1, 2, 3, 4 and 6

7.3 Knowing line graphs (3)

Q1 (a) Amber 125, Mo 134
(b) Test 4 and 6
(c) Test 5
(d) Test 2 and 3
(e) Test 1 = 47
(f) A title relating to test scores over time.

7.4 Constructing line graphs

Pupils own response.

8.1 Acute and obtuse angles

Q1 (a) Obtuse   (b) Right angle
(c) Acute   (d) Acute
(e) Obtuse

Q2

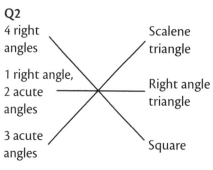

4 right angles — Scalene triangle
1 right angle, 2 acute angles — Right angle triangle
3 acute angles — Square
5 obtuse angles — Pentagon

8.2 Triangles and quadrilaterals (1)

Q1 Answers may vary, for example:

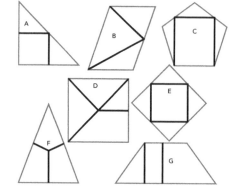

8.3 Triangles and quadrilaterals (2)

Q1 (a) Quadrilateral
(b) Triangle
(c) Quadrilateral
(d) Quadrilateral
(e) Quadrilateral
(f) Triangle

Q2

(a) Shape A = 0 acute, 0 obtuse, 4 right angles (answer may vary)
(b) Shape B = 2 acute, 1 obtuse, 0 right angle (answer may vary)
(c) Shape C = 2 acute, 1 obtuse, 1 right angle (answer may vary)
(d) Shape D = 0 acute, 0 obtuse, 4 right angles
(e) Shape E = 2 acute, 2 obtuse, 0 right angle
(f) Shape F = 2 acute, 0 obtuse, 1 right angle

8.4 Classification of triangles (1)

Q1

Right angled triangles	Obtuse-angled triangles	Acute-angled triangles
C, D, E, G	A, F	B, H

Q2 Answers may vary, for example:

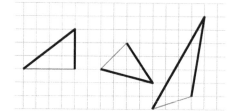

8.5 Classification of triangles (2)

Q1 Answers may vary, for example:

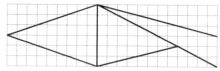

Q2

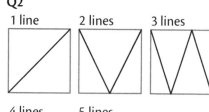

1 line   2 lines   3 lines

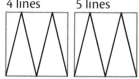

4 lines   5 lines

	Number of triangles	Type of triangles
1 line	2	Right angle
2 lines	3	2 right angle, 1 acute
3 lines	4	2 right angles, 2 acute
4 lines	5	2 right angles, 3 acute
5 lines	6	2 right angles, 4 acute

8.6 Line symmetry

Q1

Q2

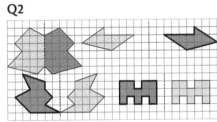

8.7 Classification of triangles (3)

Q1 (a) Scalene: 3 sides of unequal length.
(b) Isosceles: 2 sides of equal length.
(c) Equilateral: All angles and lengths of sides are the same.

Q2

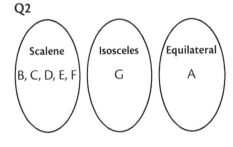

Scalene: B, C, D, E, F
Isosceles: G
Equilateral: A

8.8 Areas

Q1 (a) 20 squares (b) 22 squares
(c) 28 squares (d) 23 squares

Q2 Variable based on pupil drawings.

8.9 Areas of rectangles and squares (1)

Q1 32 cm²
72 cm²
60 cm²

Q2 (a) 81 cm²
(b) 6 cm
(c) 98 cm²
(d) Width is 6 cm and length is 12 cm

8.10 Areas of rectangles and squares (2)

Q1

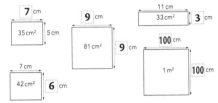

Q2 Width before = 4 cm
Width afterwards = 6 cm

8.11 Square metres

Q1

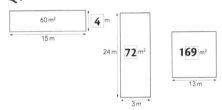

Q2 Area of the shaded part = 33 m²
Area of the shaded part = 147 m²

9.1 Converting between kilometres and metres

Q1

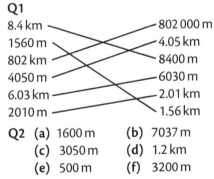

Q2 (a) 1600 m (b) 7037 m
(c) 3050 m (d) 1.2 km
(e) 500 m (f) 3200 m

9.2 Perimeters of rectangles and squares (1)

Q1 [Figures given are possibilities – pupils may provide other options that fit with the given information]

	Perimeter (cm)	Length (cm)	Width (cm)
A	40 cm	15 cm	5 cm
B	36 cm	10 cm	8 cm
C	30 cm	9 cm	6 cm
D	44 cm	11 cm	11 cm

Q2

If it was a …	Length	Width	Area
square	15 cm	15 cm	225 cm²
rectangle	20 cm	10 cm	200 cm²

9.3 Perimeters of rectangles and squares (2)

Q1 (i) 9 m
(ii) 108 m²
(iii) 66 m
(iv) Yes, vertically it would be 60 cm

Q2 Pupil responses dependent on the rectangles formed.

9.4 Perimeters and areas of rectilinear shapes

Q1 Shape A: Area = 208 cm²
Perimeter = 68 cm
Shape B: Area = 558 cm²
Perimeter = 120 cm
Shape C: Area = 504 cm²
Perimeter = 114 cm

Q2 Dependent on pupil responses. Many possibilities.

9.5 Describing positions on a 2-D grid

Q1 (a) Church (2, 2)
Supermarket (5, 6) and (6, 6)
Cinema (4, 4)
New houses (2, 4) and (2, 5) and (2, 6)
School (6, 3)
Garage (8, 2)

(b) For example: Through (3, 5) until (6, 5) and down to (6, 4)

Q2

(Grid showing: new houses, supermarket, shop, cinema, restaurant, park, park, school, church, garage — plotted on a 1–9 by 1–7 grid)

9.6 Solving problems involving time and money (1)

Q1 (a) 08:20
(b) 04:45
(c) 11:15

Q2 14 000 metres in each hour

9.7 Solving problems involving time and money (2)

Q1 (a) £7.05
(b) £0.87
(c) £2.09
(d) 1500 pence = £15.00
(e) £0.23
(f) £13.00

Q2 (a) (i) £27 a day
(ii) £4914 in half a year
(iii) £9829 in a whole year
(b) (i) £297
(ii) £2079
(iii) £33 extra a day
(iv) £231 extra a week

10.1 Calculating work rate (1)

Q1

	Child A	Child B	Child C	Child D	Child E	Child F
Questions answered	72	96	104	152	64	112
Work rate	9	12	13	19	8	14

Q2

	Baker A	Baker B	Baker C	Baker D	Baker E
Time seen	7 minutes	9 minutes	4 minutes	11 minutes	14 minutes
Cakes baked	98	108	56	66	84
Work rate	14	12	14	6	6

(a) Baker A and Baker C
(b) Baker A = 420
Baker B = 360
Baker C = 420
Baker D = 180
Baker E = 180
(c) 13 minutes
(d) They would have baked the same.

10.2 Calculating work rate (2)

Q1 (a)

2700

(b)

(12) (29) × 348

34

(c) Cyclist A cycles 18km
Cyclists B and C need to cycle 6 km an hour each.

10.3 Solving calculation questions in 3 steps (1)

Q1 (a) True
(b) False
(c) True
(d) False

Q2 Answers may vary, for example:
6 − (8 ÷ 4) − 2 = 2
4 − (6 + 2) ÷ 8 = 3
(8 + 6 + 2) ÷ 4 = 4
6 − (2 × 4 ÷ 8) = 5
(8 × 2) − 6 − 4 = 6
(6 × 2) − (8 ÷ 4) = 10

10.4 Solving calculation questions in 3 steps (2)

Q1 (a) 42 × 12 − 945 ÷ 9
42 × 12 − 105
504 − 105
399
(b) 83 + 61 × 8 − 105
83 + 488 − 105
571 − 105
466
(c) 785 ÷ 5 + 409 − 499
157 + 409 − 499
566 − 499
67
(d) 1000 + 684 ÷ 4 − 702
1000 + 171 − 702
1171 − 702
469

Q2 (a) 875 + 680 ÷ 5 − 99 = 212 (X)
875 + 136 − 99 =
1011 − 99 =
912
(b) 648 ÷ 24 + 59 × 3 = 258 (X)
27 + 59 × 3 =
27 + 177 =
204
(c) 909 + 99 ÷ 3 × 8 = 1173
909 + 33 × 8 =
909 + 264 =
1173
(d) 723 + 4 × 1554 ÷ 42 = 871
723 + 4 × 37 =
723 + 148 =
871

10.5 Solving calculation questions in 3 steps (3)

Q1 (a) Number sentence:
$88 + 62 \times 17 - 1058 = 84$
Value of a is 1054
Value of b is 1142
Value of c is 84

(b) Number sentence:
$1086 \div 6 + 26 \times 49 =$
Value of a is 1274
Value of b is 181
Value of c is 1455

(c) Number sentence:
$38 \div 38 \times 38 - 38 =$
Value of a is 1
Value of b is 38
Value of c is 0

(d) Number sentence:
$42 \times 26 \times 5 + 1005 =$
Value of a is 130
Value of b is 1135
Value of c is 47 670

Q2 (a) 516
(b) 48
(c) 2 more hours.
(d) $516 - 39 \times 12$

10.6 Solving calculation questions in 3 steps (4)

Q1 (a) $278 + [(54 \times 18) \div 6]$
First: $54 \times 18 = 972$
Next: $972 \div 6 = 162$
Then: $278 + 162 = 440$

(b) $1000 - [318 - (19 \times 14)] = 948$
First: $19 \times 14 = 266$
Next: $318 - 266 = 52$
Then: $1000 - 52 = 948$

(c) $[43 + (29 \times 23)] \div 5 \times 7 = 994$
First: $29 \times 23 = 667$
Next: $43 + 667 = 710$
Then: $710 \div 5 = 142$
After that: $142 \times 7 = 994$

(d) $551 + [(58 \times 19) \div 2] = 1102$
First: $58 \times 19 = 1102$
Next: $1102 \div 2 = 551$
Then: $551 + 551 = 1102$

Q2 (a) 13
(b) 66
(c) 2400
(d) 63 040

10.7 Working forwards

Q1 (a) $650 \xrightarrow{\div 10} 65 \xrightarrow{\times 8} 520 \xrightarrow{\div 2} 260$
Number sentence:
$650 \div 10 \times 8 \div 2$

(b) $500 \xrightarrow{\times 3} 1500 \xrightarrow{-750} 750 \xrightarrow{\div 10} 75$
Number sentence:
$(500 \times 3 - 750) \div 10$

(c) $2573 \xrightarrow{+499} 3072 \xrightarrow{-500} 2572 \xrightarrow{\div 4} 643$
Number sentence:
$(2573 + 499 - 500) \div 4$

(d) Multiply 58 by 13, then subtract 137 before dividing by 2.
Number sentence:
$(58 \times 13 - 137) \div 2 = 308.5$

Q2 (a) $53 \times 34 - (1335 \div 15) = 1713$
(b) $450 - 164 - (3 \times 90) = 16$
(c) $592 - (15 \times 30) - (592 \div 37) = 126$

10.8 Working backwards

Q1 (a) $750 \xrightarrow{\div 10} 75 \xrightarrow{\times 8} 600 \xrightarrow{\div 2} $ à 300
Number sentence:
$300 \times 2 \div 8 \times 10 = 750$

(b) $196 \xrightarrow{+579} 775 \xrightarrow{\times 3} 2325 \xrightarrow{\div 15} 155$
Number sentence:
$155 \times 15 \div 3 - 579 = 196$

(c) $3939 \xrightarrow{-999} 2940 \xrightarrow{\div 14} 210 \xrightarrow{-199} 11$
Number sentence:
$11 + 199 \times 14 + 999 = 3939$

(d) $300 \xrightarrow{\div 10} 30 \xrightarrow{\times 8} 240 \xrightarrow{\div 2} $ à 120
Number sentence:
$120 \times 2 \div 8 \times 10 = 300$

Q2 (a) $224 \div 4 - 38 \times 7 = 126$
(b) $314 \times 14 \div 7 = 628$
(c) £9670 − £1912 − £3862 − $(48 \times £4) = £3704$

10.9 Word calculation problems (1)

Q1 (a) 299
(b) 3246
(c) 126
(d) 308

Q2

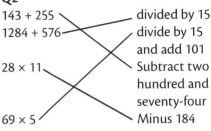

$143 + 255$ — divided by 15
$1284 + 576$ — divide by 15 and add 101
28×11 — Subtract two hundred and seventy-four
69×5 — Minus 184

10.10 Word calculation problems (2)

Q1 (a) (i) $4311 \ (498 + 3813)$
(ii) $526 \ (4339 - 3813)$ or $8152 \ (3813 + 4339)$
(b) $122 \ (61 \times 24 \div 12)$
(c) (i) $825 \ (1935 - 74 \times 150)$
(ii) $1855 \ (1935 - 2000 \div 25)$

Q2

	Number A	Number B	Number C	Number D
1	102	58	12	**1920**
2	178	**80**	18	4644
3	**207**	199	13	5278
4	1243	4295	0	**0**

Problem 1:
$(102 + 58) \times 12 = 1920$
Problem 2:
$(4644 \div 18) - 178 = 80$
Problem 3:
$(5278 \div 13) - 199 = 207$
Problem 4:
$(1243 + 4295) \times 0 = 0$

10.11 Laws of operations (1)

Q1 (a) $1256 + 295 + 173 = 1724$;
$1256 + 173 + 295 = 1724$ or
$173 + 1256 + 295 = 1724$
(b) $52 \times 16 \times 12 = 9984$;
$16 \times 52 \times 12 = 9984$ or
$12 \times 16 \times 52 = 9984$
(c) $3872 + 1243 - 751 = 5115 - 751$
(d) $A \times B \times C = AB \times C$
(e) $24 \times 67 = 67 \times 24$
(f) $99 + 99 + 99 + 99 = 198 + 198$

Q2 (a) $765 + 432 = 1197$
(b) $89 \times 288 = 25 632$
(c) $108 \times 14 = 1512$
(d) $579 + 789 = 1368$

10.12 Laws of operations (2)

Q1 (a) $120 \times 72 = 8640$
(b) $270 \times 12 = 3240$
(c) $336 \times 10 = 3360$
(d) $3136 \times 10 = 31\,360$
(e) $2500 + 940 = 3440$
(f) $125 \times 216 = 27\,000$

Q2 (a) ✓   (b) ✓
(c) ✗   (d) ✗
(e) ✗   (f) ✓
(g) ✓

10.13 Laws of operations (3)

Q1 Pupils should respond with the information that 'multiplying a number by a group of numbers added together is the same as multiplying them individually'.

Q2 (a) $45 \times 50 = 2250$
(b) $32 \times 12 = 504$
(c) $68 \times 16 = 1088$
(d) $20 \times 15 = 300$
(e) $43 \times 336 = 14448$
(f) $18 \times 8 = 144$

10.14 Laws of operations (4)

Q1 (a) 8358   (b) 4485
(c) 12 663   (d) 9831
(e) 10 241   (f) 1750

Q2 (a) $(48 + 58) \times 4 = £424$
(b) $102 \times 8 = £816$;
$51 \times 4 = £204$. Toy store A makes the most.
(c) $150\,g \times 37 = 5550\,g$ of butter;
$205\,g \times 37 = 7585\,g$ of flour

10.15 Problem solving using four operations (1)

Q1 (a) $3978 \div 26 = 153$; $3978 \div 18 = 221$; $221 - 153 = 68$ items a day
(b) (i) $1040 \div 20 = 52$ km a day
(ii) $1040 \div 10 = 104$ km a day
(c) (i) $1560 \div 60 = 26$ items a minute
(ii) $1560 \div 30 = 52$ items a minute
(d) Baker A: $1200 \div 30 = 40$;
Baker B: $1200 \div 25 = 48$;
Baker B can make 8 more cakes an hour.

10.16 Problem solving using four operations (2)

Q1 [Pupils should show 2 different methods to solve each problem
(a) 8 km long
(b) 23 pages a day
(c) No, he has only paired 29 socks in a minute
(d) $80 \div 4 = 20$ square metres a minute; 2 people can cut 40 square metres a minute; $720 \div 40 = 18$ minutes

10.17 Problem solving using four operations (3)

Q1 (a) (i) £46.25 in total
(ii) 11 weeks
(iii) Yes, £3.25
(b) (i) 112 chairs
(ii) 8 tables
(iii) 24 tables
(c) (i) £34.50
(ii) Yes, £13.50
(iii) Yes, 2 more books and 1 t-shirt, or 1 book and 2 t-shirts, or 3 t-shirts, or 3 books
(d) 22 beads

10.18 Problem solving using four operations (4)

Q1 (a) (i) 138 km in the first 2 hours
(ii) 5 hours more to complete their journey
(b) (i) £19.50 at full price
(ii) £4.50 is saved with the offer for 3 DVDs
(iii) $(3 \times £6.50) - £15 = £4.50$
(c) $216 \div 12 = 18$; $12 + 12 + 18 + 18 = $ a perimeter of 60 m
(d) (i) 80p
(ii) 60p
(iii) £2.40
(iv) £2.40

Notes

Notes